Green plants are all around us. We are totally dependent on them for food; we cultivate them for our pleasure; and we have used them in a vast number of ways down the centuries to our advantage. But have you ever wondered how plants work? Where do trees get the material to make wood? How does a bulb 'know' to sprout in the spring? Why are flowers different colors and why do they smell? This book answers these questions in a charming and accessible way. From their ability to use energy from sunlight to make their own food to their amazing range of life-sustaining, death-defying strategies, John King explains why plants dominate our planet. Plants might live life at a different pace from animals but they are just as fascinating.

This is not just a book for keen gardeners and naturalists. This is a book for anyone who wants to understand why the earth is green.

John King is Professor of Biology at the University of Saskatchewan in Canada. He was educated at Durham University in England and then for his PhD studied plant physiology at the University of Manitoba. He joined the faculty at the University of Saskatchewan in 1967 where he has remained ever since.

During a career spanning more than thirty years, Professor King has gained international recognition for his studies of plant physiology and biochemical genetics. He has served the research community in Canada for many years, especially through the Natural Sciences and Engineering Research Council in Ottawa. He is also a Past-President of the Canadian Society of Plant Physiologists and is currently an associate editor of the Canadian Journal of Botany.

Also of interest in popular science

The Thread of Life: the story of genes and genetic engineering
SUSAN ALDRIDGE

Remarkable Discoveries
FRANK ASHALL

Evolving the Mind: on the nature of matter and the origin of consciousness
A. G. CAIRNS-SMITH

Prisons of Light – black holes
KITTY FERGUSON

Extraterrestrial Intelligence
JEAN HEIDMANN

Hard to Swallow: a brief history of food
RICHARD LACEY

An Inventor in the Garden of Eden
ERIC LAITHWAITE

The Clock of Ages: why we age – how we age – winding back the clock
JOHN J. MEDINA

Improving Nature? The science and ethics of engineering
MICHAEL J. REISS AND ROGER STRAUGHAN

Beyond Science: the wider human context
JOHN POLKINGHORNE

JOHN KING

Reaching for the Sun
how plants work

PUBLISHED BY THE PRESS SYNDICATE OF THE UNIVERSITY OF CAMBRIDGE
The Pitt Building, Trumpington Street, Cambridge CB2 1RP, United Kingdom

CAMBRIDGE UNIVERSITY PRESS
The Edinburgh Building, Cambridge CB2 2RU, United Kingdom
40 West 20th Street, New York, NY 10011–4211, USA
10 Stamford Road, Oakleigh, Melbourne 3166, Australia

© Cambridge University Press 1997

This book is in copyright. Subject to statutory exception
and to the provisions of relevant collective licensing agreements,
no reproduction of any part may take place without
the written permission of Cambridge University Press.

First published 1997

Printed in the United Kingdom at the University Press, Cambridge

Typeset in Ehrhardt 11/13 pt

*A catalogue record for this book is available from
the British Library*

Library of Congress Cataloguing in Publication data

King, John, 1938–
Reaching for the sun: how plants work/John King.
p. cm.
Includes bibliographical references (p.) and index.
ISBN 0 521 55148 X (hardback). – ISBN 0 521 58738 7 (pbk.)
1. Plants. 2. Botany. I. Title.
QK50.K46 1997
571.2–dc21 96–45251 CIP

ISBN 0 521 55148 X hardback
ISBN 0 521 58738 7 paperback

Contents

Preface

The idea for this book arose from a conversation I had one day with two of my neighbors. Neither is a plant specialist but both are keen gardeners. One has in his garden a number of large trees of which not everyone is an unqualified admirer, including the other neighbor. For one thing, the trees shade adjacent gardens from direct sunlight for much of the day, including that of the second neighbor, a somewhat sensitive matter at this northerly latitude where there is a relatively short growing season. In the fall, immense numbers of leaves find their way into the general neighborhood often late in the season since some of these particular trees continue shedding leaves even after the first snow. The task of cleaning up frozen, congealed, decaying leaves is not universally appreciated.

Not for the first time, then, the owner of one of the shaded gardens was trying to persuade the tree-loving neighbor to remove his trees which, to the former, were obstacles to productive gardening. As the conversation developed it became obvious that the aggrieved party thought the main bulk of a tree came from the soil since he made repeated reference to the fact that the offending trees were taking in significant quantities of nutrients through their roots. Of course, plants do absorb many essential minerals and water from soil but we have known for a long time, for more than 300 years in fact, that air, not soil, is the source of the main building block (carbon) from which the bulk of green plants is manufactured.

This experience led me to wonder how many other aspects of the ways in which plants grow and develop are less than well known to non-specialists. Green plants (in particular, those growing on land rather than in water) have a highly specialized lifestyle due to the fact that they are generally fixed in one location throughout their lives. In addition, they contain a unique molecule, chlorophyll, which sets them apart from all other living organisms and gives them the option of manufacturing enormous quantities of their own food through photosynthesis.

Their general dependency on soil as an anchor and for essential mineral nutrients, as well as water, has led to the development of an elaborate root system with many unique functions. Like all living things, plants need water but, unlike many other organisms, they are not able to go searching for it beyond their immediate location. Thus, green plants have evolved a range of devices to obtain, transport and conserve water as well as ways to combat the effects of excessive wind, drought, cold, heat, and light from which they cannot hide.

Green plants face serious challenges due to the changing seasons. Flowering, seed production, dormancy, germination, leaf fall, and death, to name some of the more important milestones in the life cycle of plants, all are related to the seasons. Ways to measure time have evolved so that the activities of plants fit into the pattern of seasonal changes in their environment.

Green plants produce an enormous array of elaborate chemicals for only some of which a purpose is known. Plants use them to add color, fragrance, and flavor to their flowers and fruits, to wage war on predators and disease organisms, and to out-compete near neighbors. We make use of many of them ourselves, as cosmetics and pharmaceuticals for example.

These, and other features of green plants, together constitute what biologists would recognize as plant physiology – how plants work. My purpose here is to try to create an interest in and explain in straightforward language to the inquisitive layperson how plants function.

Green plants are all around us. They are the most successful of all living things evidenced by the fact that they are overwhelmingly the most abundant kinds of organisms on earth. We are absolutely dependent on them for food; we cultivate them for our pleasure; and we have used them in a vast number of ways down the centuries to our advantage. There is growing concern that we are following practices that are a serious threat to green plants. For example, we do not know what effect depletion of the ozone layer in our atmosphere will have on plants (or other creatures for that matter); the so-called 'greenhouse effect' is also an unknown quantity. We do know that we are destroying vast numbers of plants through practices such as the burning of forests, overgrazing by our domestic animals and overcropping. The advent of biotechnology and its use in agriculture is causing concern that we may be manipulating natural plant genetic processes in ways of which we are alarmingly ignorant. Overpopulation and the parallel outcome of more intensive agricultural and industrial practices (such as pollution) that go hand in hand with our burgeoning world population are a growing threat to the green plant world. Such problems should concern us all.

Here, I have tried to provide examples of some of the important aspects of how plants work with the rationale that the more people know about the lifestyle of plants, the more likely it is that they will appreciate what has to be done to preserve this component of the biosphere without which we could not survive on our earth. I also hope that people who read this will discover that knowing more about how plants work is not just useful but also fun.

I would like to acknowledge, with gratitude, the fact that much of the reading and writing for this project were accomplished during a sabbatical leave granted by the University of Saskatchewan in 1994–95. In addition, among those who helped along the way, I wish to make special mention of two; Dr. Timothy Benton, Popular Science Editor at Cambridge University Press, who gave thoughtful guidance at every step with great good humor and tact; and my wife, Myrna, who not only read and commented on each chapter but also provided ideas to help dispel the kinds of mental vapors that, surely, shroud the minds of most writers from time to time.

John King, Saskatoon.

1

Plants are cool, but why?

No doubt most of us have experienced the feeling of relief on a hot, sunny day of walking barefoot from, say, a concrete driveway or a dry, sandy beach onto a lawn. To walk even short distances across hot surfaces can be excruciatingly painful; the relief on reaching cool grass, delicious.

But why is grass so much cooler than concrete or sand? After all, the grass beside the sandy beach or the driveway is exposed to the same amount of solar energy as any other exposed surface. Yet it does not heat up to nearly the same extent. Why not?

The answer lies in the fact that plants can shed large heat loads in a number of ways, two of which are especially important. First, if the temperature of a leaf is higher than its surroundings, air circulation will remove heat from its surface by convection. Air circulation by convection depends on the fact that warm air rises because it is less dense than cooler air. The air warmed by the heat from leaf surfaces is, thus, carried up and away from the plant. As the warm air rises, it cools again, becomes more dense, and sinks back down towards the ground. This happens at leaf surfaces as long as the temperature of the leaf exceeds that of the surrounding air, as exists under extremely dry conditions, such as in hot deserts.

Alternatively, evaporation of water from leaves withdraws heat from the plant and cools it, in the same way as a room can be cooled in summer by using a fan blowing across a surface from which water is evaporating. Evaporative cooling can occur even if the temperature of the leaf is below that of the surrounding air.

Which of these two ways of losing heat (convection or evaporation) is the most important to a plant depends on the environment in which the plant is growing. For example, if a plant has a plentiful supply of water then evaporation from leaves can be great without causing damage to the plant and much heat will be lost in this way. If water is scarce, as in the example of a hot desert already mentioned, the plant has ways to conserve what small amount is available to it. Very little water evaporates from the surfaces of desert plants. Then, convective, rising air currents become the preferred method of heat removal to the atmosphere.

The tissues of desert plants, often thick and fleshy, are designed to conserve rather than evaporate water and are good examples of an adaptation to the use of air circulation to lose heat. They can often withstand temperatures above 50 degrees Celsius (°C) without suffering heat damage. On the other hand, plants such as those found in grassy lawns lose heat mainly through evaporation and feel cool to the touch. Surfaces such as dry sand or concrete are more like desert plants and rid themselves of heat mainly by air circulation, but to do this must be at a temperature above that of the air around them. Hence, they are warm, or even very hot at times, to the touch. Thus, a lawn made up of desert plants designed to conserve water would not feel much cooler than sand nearby in hot conditions. Only lawns populated by plants which are losing heat by evaporation will feel much cooler to the touch than their surroundings.

Many plants rely on their ability to evaporate water from their leaves to solve the serious problem of how to shed the solar heat they do not need and from which they cannot shield themselves. Therefore, we might expect to find that these plants are well designed for rapidly taking up water from the soil, equally rapidly transporting it through the plant, and efficiently evaporating it into the surrounding air.

But we must remember that plants also need water for their own well being, even though they evaporate away as much as 98 percent of all the water that passes through their systems in a lifetime. Plants use water directly in photosynthesis and to carry out other kinds of chemical reactions in the living cell. Water is also needed to keep living plant cells firm. This is not just a matter of keeping the whole plant rigid and preventing wilting but of maintaining the health and efficiency of cells. Water is, in fact, the most abundant constituent of living plant cells, typically 70–95 percent of the total weight of a plant. The consequences of not having sufficient water can be dire, even fatal. Thus, we would expect plants to have ways to retain water, not just absorb it from the soil, move it through the plant, and out into the air. As we shall see later, plants do have ways of conserving some of the water they take in.

During its lifetime, a plant growing in temperate regions of the world may lose the equivalent of 100 times, or indeed much more, its own weight by evaporation of water. In some tropical regions, the loss may be much higher even than this. Conversely, in hot or cold deserts, where plants tend to conserve water rather than evaporate it, the loss may be far lower.

Therefore, plants must have efficient ways to bring in water from their surroundings, usually the soil via the roots although some plants can take in water directly from the air. In tropical rainforests, for example, where the humidity is very high, aerial roots growing out from the branches are used to extract water from the atmosphere.

The cliche, 'Out of sight, out of mind', applies nowhere more appropriately than to plant roots. Because they are mostly below ground and, therefore, very difficult to observe, most people know little about roots. If we had the ability to burrow into the soil and observe the root systems of plants directly, we would be surprised at the complexity and extent of these shadowy organs.

Root systems are often enormous compared with the plant growth we see above ground. For example, the total daily increase in the length of the root system of a mature rye plant, growing under ideal conditions, was estimated at 5 kilometers (km) per day. This excluded root hairs (the delicate, threadlike extensions of the outermost layer of cells of the root just a few centimeters behind each root tip), the average daily growth of which on the same rye plant was nearly 90 km. This same rye plant had a total root length of 622 km and, in addition, the root hairs an amazing 10 620 km!

This staggering quantity of root hairs on a single plant gives some hint as to why, when plants are first transplanted, they do not grow well until they become better established in their new location. When a plant is disturbed during transplanting, many of the delicate hairs are stripped from the roots. Since hairs contribute about two-thirds to the water absorption capacity of roots, the latter will not function efficiently in water uptake from the soil until new hairs have been formed.

Roots extend continuously during a growing season into new areas of the soil searching out water and, as we shall see in chapter 5, also the minerals that are dissolved in it. This constant quest for water and nutrients takes the root system often far down into the soil. Roots at depths of 1 to 2 meters (m) are common and can be found at more than 50 m in some desert species. They often spread sideways in all directions, as well as downwards, and become very closely associated with nearly every particle of a large volume of soil from which the plant takes in the water and minerals it needs.

Once inside the root, water must be moved to all parts of the plant and especially to the leaves where it exits to the surrounding air. Some plants grow to impressive heights. For example, the giant redwood trees *(Sequoia sempervirens)* of California are tall enough, but some Douglas firs *(Pseudotsuga menziesii)* of the Pacific Northwest of North America and eucalypts of Australia are among the tallest plants in the world at about 120 to 150 m. To reach the highest leaves of such trees, water must rise from roots below the ground a vertical distance of somewhat more than this. How does a tree move water to such great heights against gravity which is tugging the water downwards?

At least a partial answer to this question was given as early as 1727, the year Stephen Hales, a versatile English clergyman and scientist, published his book entitled *Vegetable Staticks.* Hales is rightly regarded as the 'father of plant physiology' since he laid the foundation for this field of science. A good portion of his book is devoted to his experiments on the rise of water in plants which he investigated by weighing (the science of staticks) plants before, during, and after various treatments. Of the two ways plants move water over long distances, Hales did considerable work on one and hinted at the other.

Any believable explanation of the rise of water in plants must account for several important points. Just to raise water a distance of 130 m above the ground requires a push from below, or a pull from above, of 13 atmospheres. To move water upwards this distance through the dense tissues of a plant may require a force in the order of 30 atmospheres. This is a considerable pressure. By way of comparison, a car tire is inflated to a pressure of only about 2 atmospheres. A pressure of 3 atmospheres is about the point at which a diver begins to suffer the first signs of respiratory discomfort, a depth of around 30 m in water. To experience 30 atmospheres pressure, then, a diver would have to swim to a depth of around 300 m, a feat that would cause crushing problems within the body. In older terminology, but perhaps more familiar to some, the pressure at such a depth would be about 450 pounds per square inch, a formidable weight for the body to bear.

In addition to taking into account the force needed to raise water against gravity to the tops of the tallest trees, any single explanation of the rise of water up a plant must consider the speed of movement and the volume of water moved. In some hardwood trees, for example, water rises at the rate of almost 50 m per hour. A full-grown maple tree out in open country may lose more than 200 liters of water per hour on a warm sunny day. Careful measurements made under the hot conditions normally found in the mid-west of the USA showed that a maize plant may evaporate as much as 200 liters of water in a lifetime – approximately 100 times its own mass. These are high rates of movement and large volumes of fluid.

Finally, an explanation of plant water movement must be consistent with what we know to be the structure of plants. The upward flow of water in plants is known to take place in the woody tissue. Dead cells in the wood (the xylem) act as tiny tubes through which water and anything dissolved in it (the sap of the plant) moves from the roots all the way to the furthest branches and leaves. These xylem tubes form a continuous system of very narrow but open pipes running the entire length of the plant as well as into all tissues along the way. The plant, then, must have a way to force sap up through these open pipes in the xylem at a sufficiently high rate to satisfy all its needs. We might suppose, then, that somewhere in plant roots there would be a pump of some kind that would produce a pressure to force sap up a stem – what has been called *root pressure.*

The idea of root pressure is an old one that was investigated in great detail by Hales. He found that plant roots sometimes develop a pressure buildup due to their ability to absorb water rapidly from the soil and suggested that this pressure accounted for the rise of sap in a stem. The ability of plants to take in water rapidly from the soil is based on their capacity for osmosis, a process of great importance to all living organisms.

Osmosis is based on the fact that each living cell of all organisms, including plants, is surrounded by a very thin, delicate barrier, a membrane, which is so fine that it cannot be seen with the naked eye. Despite its apparent flimsiness, this barrier is of enormous importance. It not only separates what is inside each living cell from what is outside, but also controls the exchange of all the substances that are constantly moving between the interior of each living cell and its surrounding environment. The only molecule which can pass across this barrier freely in both directions is water. The strong solution of salts and other soluble compounds contained within the living cell is held in by the membrane barrier and allowed to move out only very slowly. This strong solution inside a living cell attracts water from the surroundings. This is osmosis; the attraction of water across a controlling barrier, such as a living cell membrane, in the direction of a strong solution of salts and other dissolved substances. The net result of the accumulation of water is the buildup of pressure within the living cell. The effect is similar to what happens when a balloon is filled with air. The pressure inside the balloon increases until no more air can be pumped in (without the balloon bursting, that is). In the case of the living cells of a plant root or root hairs, the buildup of pressure inside caused by the extraction of water from the soil can be used to force water from the root into the stem and the leaves.

Root pressure is well-known and significant in some plants. For example, a potent liquor called *pulque* is made by native Central Americans from the sap of the century plant *(Agave americana)*. By cutting off the single flower bud as soon as it appears, sap can be collected as it drips steadily from the cut, and then fermented. Over a period of 4 or 5 months the amount of sap seeping out under pressure from the roots of a single plant can approach 50 kg. Another familiar example is the surge of syrupy sap up the trunks of sugar maple *(Acer saccharum)* trees in the Spring, a process aided by the pressure generated by water movement in and around the roots. This sap can be tapped from the trunk of the tree in a process called 'sugaring off' and then used to produce maple syrup. Certain plants found in tropical forests pump so hard that when there is plenty of water in the roots, droplets of moisture are forced out at leaf edges.

But this root pumping is not measurable in many plants, occurs only at certain times of year in others (e.g., the sugar maple) and, even in the best examples, is not strong enough to account for the pumping of sap over the great vertical distances found in very big trees. Root pressures of even 3 atmospheres are rare, far below the kind of force needed to pump water to the top of a Douglas fir tree, for example, as we shall see later.

Even Hales recognized that root pressure could not be the full explanation for water movement within a plant. Others more recently have agreed with him. In one famous and grand experiment performed in the late nineteenth century, the German botanist, Eduard Strasburger, cut off a 20-m oak tree close to the ground and immersed the cut end of the trunk in a bath of picric acid, which kills living cells on contact. The picric acid moved into the trunk and then spread throughout the rest of the tree, killing living cells as it progressed. After several days, Strasburger replaced the picric acid with water stained with a red dye and observed that the dye eventually appeared in the leaves even though the picric acid had killed all the cells of the tree along the route used for transporting water. Thus, Strasburger showed very dramatically that a living pump mechanism was not needed to raise water to the leaves of even very tall plants. He demonstrated that even dead tissue could act as a channel through which water could be moved to great heights above ground.

A hint about what eventually became the main explanation for water transport through a plant was provided, again, by Hales. After extensive accounts of his experiments on root pressure he wrote in *Vegetable Staticks:*

> *These last experiments show, that although the* [woody tissues of the plant] *imbibe moisture plentifully; yet they have little power to protrude it farther, without the assistance of the perspiring leaves, which do greatly promote its progress.*

Thus Hales hinted that he understood the role evaporation of water from leaves played in the transport of water through the living plant which he called 'perspiration' referred to today as 'transpiration'.

The explanation which seems to satisfy all the requirements for rapid, long-distance movement of water through plants is, then, the transpiration, or evaporation, of water from leaves. The explanation depends entirely on what are known as the *cohesive* properties of water. Cohesion is a measure of the strength of the chemical bonds holding atoms and molecules together. In the case of liquid water, for example, the molecules are held together by these chemical bonds. A certain force is needed to break, or tear, the bonds apart. The amount of force that has to be applied to break the bonds is a measure of the *tensile strength* of the liquid water.

A thin column of pure water enclosed in an air-tight tube has great tensile strength. A column of pure water can withstand a pulling force of over 300 atmospheres (or 4500 pounds per square inch) without the molecules being torn apart. Among all known liquids, the tensile strength of water is by far the greatest. A tensile strength of 300 atmospheres is, for example, about one-tenth the strength of copper wire. That is, if you were to grasp the two ends of a copper wire and try to pull it apart until it snapped you would have to apply a force of some 3000 atmospheres of 'pull'.

The sap being transported around plants from the roots to the stems and leaves does not have quite as much tensile strength as pure water but has been measured certainly at 200 atmospheres. Even a thin column of plant sap, then, has the cohesion, or tensile strength, to withstand the kinds of forces needed to suck the sap up to a height of nearly 1980 m above the ground without snapping. This degree of cohesive strength is much more than enough to account for the rise of sap to the furthermost tips of the tallest trees, distances of only about 135 m.

Sap, then, is capable of being pulled, or sucked, up over great distances against the downward pull of gravity. But what creates the sucking force?

The force comes, not from a positive pumping pressure below in the roots (as we saw earlier, root pressure is not very strong even when a plant has it, and not all species do), but from the creation of a negative pressure, above. This negative pressure is similar to how we suck liquid up a straw, for example. We suck the air out of the straw first, creating a low air pressure inside the straw. Liquid then moves up the straw to replace the withdrawn air. In a similar way, water is lost from leaves as vapor by evaporation. This loss creates a low water pressure within the leaf. Replacement water is, then, simply sucked up into the leaves from the stem and roots along the tubes in the xylem that connect the leaves with the roots and, ultimately, the water supply in the soil. The tensions generated by this sucking action are large. A pull of 20 to 30 atmospheres is not unusual in very tall plants, still well within the known tensile strength of 200 atmospheres for plant sap but also well above any recorded force produced by root pressure (never larger than about 3 atmospheres). The columns of plant sap are in no danger of snapping even under such large sucking forces.

However, the large negative pressures found in the xylem tubes of plants when large amounts of water are being moved rapidly up to the leaves do create some problems. One of the most serious of these is that water put under tension is unstable. As the rate of water movement increases in response to the needs of the plant during daylight hours (mainly for cooling) tension in the water columns in the wood also increases. This increase in tension causes air dissolved in the water to escape and form gas bubbles in the tubes carrying the water up to the leaves. The formation of these bubbles breaks the continuity of the water columns and stops water transport. Too many breaks of this kind in the tube system, if not repaired, would be disastrous to the plant causing dehydration and death. Fortunately, at night, when the evaporation of water from leaves is much less than in the daytime, the tensions in the tubes decrease to the point where gas bubbles redissolve and the continuity of the columns of sap is restored.

We know ourselves how important gas bubbles in plant stems can be. Before placing cuttings (cut flowers, for example) in water we routinely remove the last few centimeters of the stem, under water of course. We do this to eliminate any air bubbles that might have found their way into the cut end of the stem and that might prevent the further uptake of water.

Plants, then, have an efficient way of transporting large volumes of water to all areas of the plant and to the surrounding air through tubes which extend all the way from the roots to the furthest tips of the highest branches. Evaporation of water as vapor from leaf surfaces causes more liquid water to be sucked up into the plant through the roots from the soil. In this way, the entire plant is supplied with water and the minerals dissolved in it, all of which are essential to the continued well-being of the growing plant. As well, the evaporation of water from the leaf surfaces helps keep the plant cool and functional even during very hot weather.

But it would be wrong to end here and leave the impression that the transport of water through a plant and out through leaves is an isolated process. Hales in his *Vegetable Staticks* made reference to the other major role that leaves play in the life of the plant: photosynthesis.

> *leaves seem also designed for many noble and important services . . . plants probably drawing thro' their leaves some part of their nourishment from the air.*

Thus Hales hinted at the fact that plants growing on land (as opposed to spending their entire lives in water) are faced with a major dilemma. They are surrounded by a dry atmosphere into which they evaporate water vapor, a useful process as we have seen. However, the air is so far from being saturated with water that the plant would be in danger of lethal dehydration if the loss of water vapor through leaves were not somehow controlled.

Yet, at the same time, plants must take in from the air the gases they need for respiration (oxygen) and for photosynthesis (carbon dioxide). As these gases enter the leaf, water vapor is lost. But what if the plant is short of water? How can the loss of water vapor be stopped or at least slowed, without harming the plant by cutting off the flow of oxygen and carbon dioxide into the leaf? This is the dilemma land plants inevitably face in their life cycle.

The short answer is that the plant compromises in balancing these competing requirements. In the first place, most of the above-ground surface of the plant is covered with a thin layer of wax which gases and water cannot penetrate. We can see this waxy barrier (the *cuticle*) in many leaves which appear shiny and are slippery to the touch. This cuticle covering the plant surfaces exposed to the atmosphere serves as an effective barrier to water loss and protects the plant from desiccation as well as from attack by insects and other predators. However, a complete barrier to water vapor loss would also block intake of oxygen and carbon dioxide. How does the plant allow this barrier to be breached? The answer to this question is, in much the same way as insects do. Insects have an outer cuticle that is impermeable to air. They, too, are faced with the problem of exchange of air in breathing. They overcome the problem by having breathing pores, holes through which gases can be exchanged with the atmosphere.

Usually on the underside of their leaves, plants also have breathing pores called stomata (the word is Greek for mouths). Each stoma has two cells, one on either side of its opening, known as guard cells, that are shaped rather like the lips of a mouth. The two guard cells are designed so that when they are filled to capacity with as much water as they will hold, they separate from one another. Through the gap that appears between them gases and vapors can be exchanged between the interior of the leaf and the surrounding air, just as happens when our lips open. At other times, the guard cells lose water again, come together (as we might close our lips), and the stomata close. Then gas exchange with the atmosphere is cut off. Control of the movement of water into and out of the guard cells themselves is by osmosis, a process described earlier in this chapter in relation to root pressure.

The number of stomata on the underside of a leaf is normally so great that if they were always open the leaf would evaporate most of its water. The ability to open and shut, however, allows control of water loss. Guard cells function as valves with multiple controls to open and close them. Changes in such things as the brightness of sunlight, temperature and relative humidity in their surroundings, and carbon dioxide concentrations inside the leaf, are all somehow sensed by the guard cells which react accordingly by changing their shape, moving water in and out through osmosis, and, hence, altering the size of the opening between them. At night, when there is no photosynthesis and thus no demand for carbon dioxide within the leaf, stomatal openings are kept small, preventing unnecessary loss of water since the heat load on the plant is also small at night and evaporation is not necessary for cooling. During the early part of a sunny day the demand inside the leaf for carbon dioxide for photosynthesis is large and the stomatal pores are wide open allowing free exchange of gases.

Water vapor loss is also great under conditions where the stomata are wide open, but as long as water is plentiful it is to the advantage of the plant to trade water for food. When water is not so abundant, the stomata will remain partially, or even completely, closed so that the plant will avoid potentially lethal dehydration. Of course then, the ability to produce food substances through photosynthesis will be severely hindered.

But a plant can survive for much longer without a source of food than without water, just as we can. When was the last time you heard of someone going on a water strike? Hunger strikes are common means of protest but not water strikes. We can live for weeks without food but not many days without water, even in the best cool conditions. In a hot, dry desert, lack of water can be fatal in a matter of hours. A plant is much the same as an animal and will conserve water at the cost of making new food through photosynthesis. The best way, then, to encourage productivity in our gardens is to keep crops well watered so that the stomata in leaves will be kept wide open for photosynthesis.

We began this chapter by exploring the notion that many plants keep cool by evaporating water from their leaves. In the course of learning how this happens we have also found out that much more is involved in how water is used by plants. Water is essential to all life. Plants use it in many ways other than in keeping cool and have evolved ways to take it in and move it, often over great distances, against gravity with high efficiency.

Yes, plants are cool, thanks to their smooth systems for handling large quantities of water. But their coolness is only one aspect of their *sangfroid!*

2

Photosynthesis: the *leitmotiv* of life

Those who answer gardening questions from the general public in newspapers, on radio, and the like will tell you that surprising numbers of people have a basic misconception about plants. The belief that plants somehow build themselves from the soil is widespread even today, more than 300 years after excellent proof was gathered showing this not to be so. Why the belief still exists at all is something of a puzzle. Consider the common practice of bagging lawn clippings. If grass was simply built from soil, a lawn from which several pounds of clippings were removed every week during the growing season for the last dozen years would likely resemble a sunken garden by now. Obviously, something other than just the soil must go into the building of a plant.

As unlikely as it may seem at first, green plants, in fact, make themselves out of carbon dioxide, water and minerals, with the aid of light energy. What green plants make by this *photosynthesis* (literally, putting together by light) is an endless supply of carbohydrates, such as sugars (like glucose and sucrose), starch, and cellulose, the most common sources of foods and fuels available to us and other animals.

Not that plants are the only living organisms capable of this feat. For example our oceans, lakes, and rivers are populated by other kinds of living things also capable of photosynthesis. These belong to a large group of organisms called the protists. Familiar to us in this large grouping of living things are the green algae we often see at certain times of year as 'scum' on ponds and in lakes; the larger green, brown, and red algae, or seaweeds, on seashores; and microscopic organisms, like the diatoms, dinoflagellates, and the smaller algae, which make up the plankton so crucial as food sources for other marine and freshwater wildlife. All of these protists can also build themselves from simple substances in their environment, with the aid of light. Here, though, I shall emphasize plants since they are more familiar and accessible to us.

How plants and protists construct themselves out of carbon dioxide, water, and a few minerals is an intriguing problem and one that should be endlessly fascinating to us for very practical reasons. Our oil resources originate from carbohydrates formed many millions of years ago, mainly by ancestors of modern day protists growing in the oceans of the earth. Ancient vegetation (mainly cellulose) is the source of our coal reserves. Green plants are also the ultimate source of the foods that fuel the great majority of living organisms. The most common of these foods are the carbohydrates which are formed directly during photosynthesis.

All of these interests have led to attempts going back several centuries to understand in more and more detail how green plants, primarily, and protists capture light energy, use it to release hydrogen and oxygen from water, and then use the hydrogen together with carbon dioxide to form sugars and other carbohydrates. Some hope that eventually we may be able to exploit photosynthesis artificially for our own fuel needs. If we could split water into its component atoms, hydrogen and oxygen, using light as a cheap energy source, we would have a supply of chemical energy in the form of hydrogen from an unlimited, endlessly renewable resource. Understanding photosynthesis, then, is not just a matter of curiosity about how green plants do something so apparently miraculous with such ease but is important because there are possible practical implications as our coal, oil, and gas reserves, themselves the result of photosynthesis long ago, are depleted.

Questions about energy form the heart of investigations into photosynthesis; how green plants can use light energy from the sun and convert it into carbohydrates. Living things need energy to live. Energy drives all that living things do. We all understand the need for energy to do physical work. Less obviously, we also need it for our brains to plan, or hope, or dream during sleep. In fact, it takes more energy to power our brains even while we are asleep than it does to light a 60-watt bulb in our bedside lamp.

For animals, energy comes from the food they eat. But in fact, all living things must have an external energy or fuel source. This is because energy is needed to run every function of all living things from the smallest microbes to the largest animals and plants. The burning of fuel, whether it be in a car, a light bulb, a computer microchip, or a living organism, always leads to the production of heat as a by-product. In fact, on average, about 90 percent of the energy in any source of fuel ends up as heat, eventually, as the fuel is used. We may not always be aware of this heat because most organisms are not warm to the touch as we are. In warm-blooded animals like ourselves, the production of heat is obvious. What is less clear is that heat loss occurs in all living organisms whether warm to the touch or not. Thus, we all need fuel or food, if you prefer, to replace the energy we lose as heat. What are these food sources and where do they come from? Before answering that question, we need to know a little more about the nature of the fuels we are talking about.

One of the earliest explanations of the use by living things of fuel as a source of energy came from the eighteenth century French chemist, Antoine Lavoisier, who wrote:

> *Respiration* [breathing] *is merely a slow combustion* [burning] *of carbon and hydrogen, which is similar in every respect to that which occurs in a lighted lamp or candle, and, from this point of view, animals that breathe are really combustible bodies which are consumed.*

Lavoisier's point was that animals had in them fuels containing carbon and hydrogen which could be slowly 'burned' in respiration releasing energy. This brilliant insight unfortunately did Lavoisier absolutely no good. He was guillotined during the French Revolution, the judge reportedly dismissing him with the sentiment, 'The Republic has no need of savants [men of learning]'.

Today, we understand, in a way not possible in Lavoisier's time, that nearly all the thousands of natural chemicals found in living things contain carbon and hydrogen. These biological molecules are called *organic compounds,* because at one time it was thought that chemical compounds containing carbon were produced *only* by living organisms. We know today that this is not so. Over five million different organic compounds are now known. Many of them are artificial and are quite unlike any of the natural compounds produced in living organisms although they are still called 'organic' because they contain carbon. No other chemical element comes close to matching the diversity of the compounds that carbon forms.

Among the natural organic compounds found in living organisms are the carbohydrates which are made up of carbon, hydrogen, and oxygen. Some of these are among the most important fuels burned, as Lavoisier suggested, in respiration to release energy. Carbohydrates familiar to us are sugars like glucose (grape sugar) and sucrose (table sugar), as well as starch and cellulose. Foods containing sucrose or starch are well-known fuels. Cellulose is the very strong fiber that the plant's basic structure is made of (not to be confused with cellulite, the fat we can have vacuumed from our thighs if we are sufficiently wealthy). The page of paper you are now reading is made of cellulose fibers produced by trees and used as wood pulp.

Cellulose as a fuel is less familiar since we, along with most other animals, cannot digest it. The fiber is useless to us, therefore, as a *direct* source of energy. Only ruminant animals, such as cattle, sheep, and deer, and some other organisms, like horses and termites, have adapted to the indirect use of cellulose as a fuel. They all have bacteria in their digestive systems to break down cellulose into a form which the bacteria can use as fuel. The animals harboring the microbes benefit by stealing the surplus fuel not needed by the bacteria in return for providing a favorable environment in which these particular organisms thrive.

Before going further, though, we need to digress for a moment and say a couple of things about the carbohydrates, sucrose, starch, and cellulose. Most importantly, all three are produced in green plants during photosynthesis. Then, when living organisms use them as fuel, they are first broken down to glucose. In fact, starch and cellulose are simply made up of many hundreds of glucose molecules linked together like beads on a chain. When used as fuel by living organisms these long chains are simply broken up to release the individual beads, the glucose molecules, already there. Thus, glucose is the common denominator fuel whether sucrose, starch, or cellulose was the source.

Crops such as sugarcane and sugarbeet are major sources of sucrose although we should remember that many of the foods we eat, fruits being other examples, are also packed with sugars. I hope nobody imagines they can lose weight by replacing meat with fruit in their diet! Starch is the main fuel found in many of our major staple food crops like wheat, rice, maize, sorghum, and potatoes. It is also the main fuel stored in bulbs, corms, tubers, and rhizomes. And cellulose is everywhere in green plants. In fact, it is not so far wrong to say that plants *are* cellulose. Leaves, stems, and roots are built mainly of it. When we admire a lovely green landscape or contemplate the trees of the forest, what we see is mainly cellulose, to reduce the beauty of nature to something rather less poetic!

So carbohydrates are wonderful fuels, excellent sources of energy, and are produced in abundance by green plants. We certainly have some idea of how much energy cellulose contains when we burn logs on a fire and feel the intense heat given off. Table sugar will also flare up dangerously if a handful of it is thrown on an open flame, demonstrating that it, too, contains a large amount of energy.

But, as pointed out already, at the center of the use of carbohydrates as fuels is glucose. It is the glucose released from carbohydrates like sucrose, starch, or cellulose, that is burned, with the help of oxygen from the air, in respiration, to yield carbon dioxide, water, and a great deal of energy (see chapter 3). The energy released is then used by organisms to power their own systems.

Coming back to photosynthesis, the question we need to ask at this point is, how does the reverse process occur? We can understand how sucrose, starch and cellulose are used to produce glucose which then can be burned in respiration to release an enormous amount of energy. But how does the reverse happen? Making glucose from carbon dioxide and water takes just as much energy as is released in the burning of the carbohydrates in respiration – a great deal of energy! The by-products of respiration, carbon dioxide, and water, are extremely stable, unreactive chemical compounds. Therefore, how does the production of glucose from these compounds happen?

There are lots of carbon dioxide and water molecules in the air, oceans, lakes, the soil, and inside living organisms, but the chances of any of them simply coming together in the right way to produce even a single molecule of glucose are extremely remote. Such a thing is not likely to have happened in the billions of years carbon dioxide and water have existed side by side on the earth. The reason for this is because of the need to supply large amounts of energy to force carbon dioxide and water to combine to form glucose and release oxygen gas. The chemical bonds holding hydrogen and oxygen together in water molecules, and carbon to oxygen in carbon dioxide, are extremely strong. Such tightly locked structures can be broken apart only by applying a great deal of force. But, they must be broken up if the carbon, hydrogen, and oxygen of carbon dioxide and water are to interact to produce glucose.

Yet, green plants form glucose from carbon dioxide and water every daylight hour during their growing seasons. Having accomplished that impressive feat, plants then go on to produce a seemingly endless supply of sucrose, starch, and cellulose from the glucose. How they do all this is the story of *photosynthesis,* the process which distinguishes green plants and protists from all other organisms. To appreciate more fully exactly how plants do it we need to know some of the history behind our understanding of photosynthesis.

Although, as we saw earlier in this chapter, Lavoisier was one of the first to study animal respiration and to understand the need animals have for energy, it was known before his day that animals must eat to live.

Until some 350 years ago, it was thought that plants obtained their food in a similar way to animals, that is by 'eating', in the case of plants, whole matter from the soil. After all, at the time, people believed that all things consisted of 'the four elements', earth, air, fire, and water. Air and fire were recognized as having no weight and so it was thought that anything with weight could only come from earth, water, or both.

These early ideas began to alter dramatically during the seventeenth century. At the forefront of change was Jan Baptista van Helmont, a Belgian physician. In one of the first recorded scientific experiments, Van Helmont observed and measured the growth of a willow tree. He weighed the tree at the beginning and again at the end of a 5-year period and found that it gained nearly 75 kg. Yet, the soil in the pot in which the plant was grown lost only a few grams over the same time period. Van Helmont was forced to conclude that the body of the willow could not possibly all have come from the earth in which the plant was grown. In fact, it seemed that most of it had come from elsewhere since the soil lost only a fraction of the weight gained by the plant.

But then van Helmont made an error. He reasoned that if most of the weight gained by the willow was not from the soil then it must be from the water added to the tree during the 5 years of the experiment. Unfortunately, he, also, was the victim of the primitive state of knowledge of the natural world at the time. As the elements water and earth were thought to be the only sources of bulk, it followed that if earth (soil) was not the source of the gain in weight of the willow, then it had to be the water added, since the elements air and fire had no weight. As we shall see later, van Helmont was not wrong in saying that water contributed to the weight gain by his willow tree. His mistake was thinking that water was the *main* origin of it. What was particularly startling to everyone at the time was that the soil itself contributed so *little* to the weight gain. This inescapable conclusion from van Helmont's experiment completely changed the thinking about where plant food came from.

There was no possibility of understanding van Helmont's measurements until knowledge of chemical elements improved. Progress had to be made in thinking beyond the four elements. At the forefront of these advances, along with others like Lavoisier, was the late eighteenth century English clergyman and chemist, Joseph Priestley.

Combustion, or burning, was a topic that had intrigued alchemists for centuries, especially how one chemical compound could be transmuted into another by heat. The age-old search by alchemists for the formulae to transmute baser metals into gold and for the elixir of life were manifestations of the fascination they had with the effects of combustion on substances. Modern-day industries based on the purification of metals and their fusion together by heating to form alloys are outgrowths of the activities of these alchemists. The successors to the alchemists, chemists such as Priestley and Lavoisier, were equally intrigued by combustion, as the quotation earlier in the chapter from Lavoisier indicates.

Priestley was fascinated by the fact that combustion somehow 'injured' air. He performed some bizarre experiments to demonstrate this 'injury'. For example, if he burned a candle in an air-tight container, the flame soon went out. If he then put a mouse into the container, the animal died, because the air in the container was said to have been injured by the burning of the candle. Priestley found, however, that if he put a sprig of fresh mint into an air-tight vessel containing injured air, the air was eventually restored to a state in which it would:

> *neither extinguish a candle nor was it at all inconvenient to a mouse, which I had put into it.*

In other words, the mouse did not die, was not 'inconvenienced', as Priestley so delicately put it, as long as the air in the container was 'restored' after a candle had burned out in it.

Priestley's conclusion was that vegetation restored air by cleansing and purifying it thus removing the injury. What he did not understand was that these restorative powers of mint, as well as the leaves of other plants, as he found later, depended on the presence of light. He used glass containers in his observations so that he could see what was going on inside. In so doing, he accidentally introduced light into his experiments but never did figure out the significance of it.

The importance of light in the restoration of injured air was left to the Dutch physician, Jan Ingen-Housz, to discover a few years after Priestley. Ingen-Housz found air to be purified by vegetation only in sunlight. In addition, only the green parts of plants had the 'power to mend air'. Ingen-Housz went even further by finding that plants absorbed the carbon in carbon dioxide:

> *throwing out at that time the oxygen alone, and keeping the carbon to itself as nourishment.*

Here was the first hint that light and carbon dioxide were linked in some way leading to the release of oxygen and that oxygen was the restorative in injured air. Here also was the first hint that air provided a chemical element, carbon, essential for the nourishment of a plant. These and other, similar, results made it clear that there was more to air than had been imagined. Perhaps it was, after all, possible to 'gain weight' even from something as insubstantial as air.

One other essential observation was needed before this 'synthesis by light' could be fully understood. The Swiss scientist, Nicholas Theodore de Saussure, made the final connection in the first decade of the nineteenth century. He showed that in the light a plant released exactly as much oxygen as its uptake of carbon dioxide. More importantly, he also showed that the weight a plant gained was greater than could be explained by the amount of carbon taken in as carbon dioxide. In other words, in addition to carbon dioxide, something else contributed to the solid substance of a plant. This extra contribution, de Saussure showed, had to be from water, since he was able to eliminate all other possibilities in his investigations.

So, van Helmont was not so wrong after all. Water *is* critical to weight gain in a plant but is not the main source of the increase, as van Helmont had thought. Taken together, all of these early studies made it clear that the main bulk of plants could, after all, be produced largely from something as apparently weightless as air, in light only and with the secondary help of water. The realization of the importance of carbon dioxide from the air and the need for sunlight for weight gain in plants shifted attention in the study of photosynthesis away from the soil to where it belonged; to the role played by the atmosphere and by light. Green plants will flourish, we now know, if supplied with air containing carbon dioxide, with water and minerals from the soil, and with sunlight. When sunlight strikes a healthy green plant, the plant uses the energy of the light to capture carbon dioxide from the air. This carbon dioxide is then used as a building block in the construction of sugars and other carbohydrates. These become the plant's only source of food or fuel.

We all can appreciate that there is no lack of energy in sunlight. The heat of summer and sunburn remind us of that fact, regularly. The amount of solar radiation reaching the surface of the earth is about 4 exajoules per year (an exajoule is one billion – or one thousand million – joules of energy, approximately equivalent to the amount of heat released in the burning of 22 million tonnes of oil). Put another way, the amount of solar energy falling on each one and one-half square miles of the earth on a bright, sunny summer day is equivalent, on average, to the power of an atomic bomb of the Hiroshima type. About 60 percent of this enormous amount of solar energy is reflected directly back into space (recall the images of the brilliant, shining Planet Earth as seen from the surface of the moon by those who landed there during the 1970s). The brightness of the earth is an indication of the amount of light from the sun reflected back into space. Most of the rest of the solar energy is absorbed by the atmosphere, by clouds, or by oceans and landmasses, and promptly radiated back into space as heat. On this scale of the earth's energy budget, the amount of sunlight absorbed by green plants and used in photosynthesis is tiny; less than 1 percent even of the fraction of solar energy that penetrates to the earth's surface.

Only when light is absorbed by something can its energy be used. Since the late nineteenth century we have known that when light is absorbed by metals, for example, electrons are dislodged from the metal. If collected in an efficient way, these electrons can be organized into an electric current (electricity is just a stream of electrons moving along a wire). In fact, some kinds of burglar alarms, photographic exposure meters, television cameras, solar-powered calculators, and the solar panels on orbiting earth satellites all operate on the principle of turning light energy into electrical current.

In a plant, sunlight is not absorbed by metals but by *pigments,* molecules whose bright colors signify that they strongly absorb only some of the wavelengths of light from the sun. The most important pigment found in plant leaves is chlorophyll which absorbs blue and red wavelengths of light. The wavelengths not absorbed, those that are reflected from the leaf or just pass right through it, give the pigment its characteristic green color to our eyes. Several pigments in addition to chlorophyll are found in most plant leaves but cannot be seen, because of the predominance of green pigment, until the fall when chlorophyll disappears from leaves. All of these other pigments absorb bluish-violet light strongly so we see them as being yellow, orange or red, depending on exactly which wavelengths they absorb (see chapter 12 for a discussion of plant pigments and color).

Just as in the case of metals, the absorption of light by these pigments, but principally by chlorophyll, dislodges electrons from the pigment molecules. These electrons are then organized within leaf cells into tiny electric currents. This is the energy from light that is then used by the plant to power a complicated sequence of chemical reactions that has the net effect of splitting water molecules into hydrogen and oxygen and then transferring the hydrogen to molecules of carbon dioxide. The outcome of this sequence is the release from the leaf into the atmosphere of oxygen from the water and the formation of glucose from the combination of carbon dioxide and hydrogen.

The first carbohydrate formed in photosynthesis may be glucose but this is just the beginning. Some of the glucose is used as fuel to supply energy to the plant for its own life functions just as happens in all living organisms. But, during most days there is light energy enough for the leaves to produce much more glucose than the plant requires for its own immediate needs. Surplus glucose is diverted to the production of storage carbohydrates, such as sucrose or starch, and to the manufacture of cellulose fibers for building the plant structure.

So, there is no doubt that photosynthesis is absolutely vital to the growing plant. Its immediate food supply for respiration, its stores of fuel for the future, and its very structure all come from the glucose formed in photosynthesis. Plants are found everywhere, from arctic tundra to tropical forests to arid deserts. It is not surprising, then, that if we look at a wide range of species we find clear evidence of how green plants have evolved and adapted to make this most critical of all life processes, photosynthesis, as efficient as possible in an enormously wide range of global environments.

The leaf is where we find most of the photosynthetic activity going on in most plants. Not surprisingly, then, we find that the leaf is beautifully designed for this crucial function. Leaves can vary enormously in size and shape but the basic leaf is flat and thin. This gives to a leaf a maximum surface area together with a minimum volume, a design that allows the leaf the greatest area on which to receive maximum light and carbon dioxide from its surroundings with no waste of thickness. The layering of the leaves in the canopy of the plant maximizes the exposure of the greatest number of leaves to sunlight. The leaves of some plants are also capable of adjusting their position to follow the sun as it moves across the sky, a process called *solar tracking.* All these adaptations aid photosynthesis.

The interior of a leaf is also highly specialized for light absorption. The region of the leaf just below its upper surface is made up of *palisade cells,* so named because they are narrow and tall like the individual boards in a palisade fence. They stand in tightly packed, parallel rows one to three layers deep. We may wonder how efficient it is for a plant to develop more than one layer of these cells when we might imagine that the first layer would absorb most of the sunlight, anyway. In fact, more light than might be expected goes through the first palisade layer because of, what are called, *sieve* and *light guide* effects.

The sieve effect is caused by the fact that chlorophyll, the main pigment that absorbs light in photosynthesis, is not evenly distributed within leaf cells. To the naked eye a leaf looks uniformly green. But, if we magnify the inside of a leaf using a microscope we find that green chlorophyll is not spread evenly throughout every cell in the leaf. Under the microscope we find leaf cells have in them what look like many tiny green packages, called chloroplasts. The remaining contents of each cell are clear, not green at all. This clustering of the pigments of photosynthesis in chloroplasts means that not all the chlorophyll molecules in each chloroplast are equally exposed to sunlight. Some molecules are shaded by others. Not every chlorophyll molecule, therefore, has the same opportunity to absorb light as its companions in a chloroplast. In addition, some light passes in between the chloroplasts through the clear areas of a cell and is not intercepted at all. Thus, one way or another, some light is not absorbed by the first layer of palisade cells and travels on to the second and third layers in the sieve effect.

The light guide effect refers to the channeling of some light between the palisade cells, deep into the leaf. Because the palisade cells are tightly packed and arranged in such orderly, regular layers, the narrow spaces between the cells can act in the same manner as an optical fiber. In fact, plants have the ability to guide light in this way several centimeters into their tissues. Thus, in some smaller seedlings, these 'light pipes' may direct light all the way down into the roots, which may not always be in the total darkness we assume them to be. In the case of leaves, some of the light striking the upper surface of a leaf is directed down between the palisade cells via the light pipes. Light is channeled in this way deep into the interior, towards the lower half of the leaf where a quite different arrangement of cells is found.

Below the highly organized palisade layer, is so-called, *spongy* tissue where the cells are very loosely packed, unlike the tight packing in the palisade tissue. Cells in the spongy tissue are irregularly shaped and, because of that, have large air spaces between them. Such an arrangement produces many surfaces from which light can bounce and be scattered. The light channeled between the palisade cells bounces around among these surfaces in the spongy tissue, scattering in all directions. We can see something of the effect of this scattering with the naked eye. Look at the lower surface of many kinds of leaves and you will find them to be lighter green than the upper surfaces. Some of this lighter shade is caused by light scattering from cell surfaces in the spongy layer. Bouncing around the light in this way gives the cells towards the lower, shadier side of a leaf more opportunities to absorb some of it for use in photosynthesis.

The contrasting arrangement of cells in the palisade and spongy layers – the former allowing some light to pass through and the latter trapping as much light as possible – allows a leaf to intercept and use in photosynthesis a maximum amount of the sunlight entering from above.

But capturing light as efficiently as possible is only one of the problems faced by the photosynthesizing leaf. Equally important is the problem of capturing carbon dioxide from the air and moving it to where photosynthesis is occurring inside the leaf. The main problem here is the small amount of carbon dioxide in our atmosphere to begin with, even though we are causing it to increase as fast as we seem able by burning fossil fuels at a scandalous speed.

Oxygen is at a concentration of about 210 000 parts per million in air. Since the dawn of the industrial age, the concentration of carbon dioxide in the atmosphere has increased from about 280 to 350 parts per million. So, compared to oxygen, there is really not much carbon dioxide available to a plant. Anything that improves the efficiency of a leaf to take in carbon dioxide from the small amount available in the atmosphere will be favorable to photosynthesis, therefore.

Carbon dioxide enters a plant through the stomata, the pores usually found on the undersurface of a leaf (see chapter 1). The gas then finds its way to the cells of the spongy and palisade layers of the leaf where photosynthesis is occurring. In the case of more than 95 percent of all plant species, carbon dioxide is simply taken into leaves through the stomata and, if light is available, used directly in photosynthesis. There are thousands of stomata in each leaf and they remain open all day as long as the plant is well supplied with water (see chapter 1).

About 3 percent of plant species have a different, two-phase system of photosynthesis. These species have a first phase in which a carbon dioxide pump sucks the gas more efficiently out of the air into the leaf. Of course, the pump is chemical rather than mechanical, as in the case of a water or oil pump for example. The leaves of these plants have a chemical way of capturing carbon dioxide very efficiently and then, in a second phase, releasing it again inside the leaf to the cells when and where it is needed for photosynthesis in the same way as in all other plants. Although only about 3 percent of known plant species have such pumps, some of them are among the most important food plants in the world; such as maize, sorghum, and sugarcane. The question is, why did certain plants develop a two-phase photosynthesis of this kind? The answer is that it seems to depend on where the plants originated.

Plants like sugarcane and maize seem to have originated in tropical grasslands. These are far from being deserts but they are environments where there are long periods of hot, dry weather every year. During prolonged periods of heat and drought, the stomata of plants in these areas are often either completely or partially closed for much of the day to conserve water. Consequently, the ability of these plants to take in carbon dioxide through their stomata is severely limited for long periods. Growth would be severely impaired if some way were not found to pump carbon dioxide into leaves so that photosynthesis could continue.

To help overcome this difficulty, certain plants have developed a way of sucking carbon dioxide into their leaves very efficiently, even through partially closed stomata, and then delivering it in high concentration to locations where photosynthesis is occurring. All of this aids plants growing under conditions of high temperature and restricted water. An added bonus is that the partially closed stomata limit the loss of water. As we saw in chapter 1, this is also of crucial importance to plants.

A more extreme example of a carbon dioxide pump can be found in the cacti and many other kinds of fleshy plants found in deserts. Here, the shortage of water is often so acute that stomata remain closed all day. Many desert plants minimize water loss by opening stomata only at night. But if stomata are closed all day, and photosynthesis can go on only during daylight hours, how is the problem of capturing the necessary carbon dioxide solved?

Many desert species have solved this problem by using a chemical carbon dioxide pump similar to the one found in plants growing in tropical grasslands. During the night desert plants open their stomata and take in carbon dioxide which they simply store. Throughout the following day, they release the gas once more slowly from this storage, behind closed stomata, to where it is needed for photosynthesis. Pineapple is the best known food plant able to perform this trick.

The realm of living things is entirely distinct from the physical world in organization and how it works. And yet, the organic world is entirely dependent on the inorganic for the means to maintain its life processes. Important among these means is the need for carbon and hydrogen in forms which will serve as fuel to burn with oxygen to release energy. But in the inorganic world, carbon and hydrogen are locked up in two of the most stable, unreactive substances known, carbon dioxide and water.

It is here that plants, but not forgetting other photosynthetic organisms, play their key role in the drama of life. Plants are capable of capturing light and converting it to chemical energy. The plant uses this energy to reshuffle the atoms in carbon dioxide and water to make carbohydrates, the most basic of all foods. At the same time, the plant releases oxygen, an essential ingredient for the eventual burning of foods to supply the energy needs of all organisms, as we see in the next chapter. In fact, plants are nothing if not carbohydrate factories, as we have seen. Little wonder, then, that the question of how plants carry out photosynthesis has been one of our preoccupations for a very long time. Its importance to all life ensures that photosynthesis will also continue to occupy our attention far into the future.

3

Respiration: breathing without lungs

Arguably, one of the more impressive achievements in biology in the twentieth century has been the gaining of a deeper and clearer understanding of the many steps and processes involved in what we call respiration. The knowledge most people have of respiration often begins and ends with the fact that they know it to involve the intake of oxygen and the release of carbon dioxide since this is what happens when they inhale and exhale – we breathe in air for the oxygen it contains and breathe out the air again enriched with the carbon dioxide we produce but do not want and cannot use.

What, in fact, is going on inside us is that the foods we eat, carbohydrates, fats, proteins and other substances, are being slowly combusted or burned as Lavoisier described it more than 250 years ago (see chapter 2). Using the oxygen of the air we breathe in, we slowly burn the foods we take in, converting them eventually to carbon dioxide and water. At the same time, the energy contained in the foods is released, some of it in a form that is useful to us; the rest is simply given off to our surroundings as heat. We then take the useful energy and put it to work to sustain our life support systems – to drive our muscles and other organs, keep us warm, feed our brains, and build the complex molecules that together constitute what we call 'life'.

This process of respiration so characteristic of animals like ourselves is equally characteristic of plants. We might suppose that because plants have access to the enormous quantity of energy in sunlight through photosynthesis they would not need to have any additional supply from any other source. This is not the case, however. For one thing, not all parts of a plant carry out photosynthesis, only those that are green. The non-green parts of plants also need energy. In addition, the capture of light energy in photosynthesis can only go on during daylight hours but plants grow all the time, day and night, as long as they are in a favorable environment. Energy is required 24 hours a day, not just when it is light. Reliance on light energy is not sufficient for these continuous needs. Respiration provides the additional energy needed to build the many types of molecules that plants need for their life processes. Fats, proteins, and nucleic acids are obvious ones that plants have in common with other organisms. In addition, though, plants produce a vast array of rather exotic molecules that are specially designated for such things as defense, as attractants and for protection from the environment (see chapters 11 to 16).

Our knowledge of respiration in plants had its beginning in the seventeenth century when it was discovered that seeds needed to be exposed to air before they would germinate. However, it was not until the work of Lavoisier, Priestley, and others about a century later (see chapter 2) that the exchange of gases, like oxygen and carbon dioxide, between organisms and their surroundings began to be appreciated. Until that time, what air was and what it contained were largely unknown. Air was insubstantial, invisible and rather mysterious.

By the last quarter of the eighteenth century, it was recognized that germinating seeds took in oxygen and gave out carbon dioxide, as Lavoisier, Priestley, and others had found animals did when they breathed. By the late eighteenth century, Ingen-Housz had shown that all living plants gave out carbon dioxide in the dark, and that non-green plants did so in the light as well. What green plants did in the light was much more difficult to figure out, as we shall see shortly.

It was left to the Swiss scientist, de Saussure, to begin the really detailed investigation of plant respiration. In the last decade of the eighteenth century, de Saussure showed that plants and animals seemed to produce and release carbon dioxide in very much the same way and in similar quantities. He went on to show that all parts of a plant (leaves, flowers, fruits, roots, etc.) exchanged oxygen and carbon dioxide with the surrounding air.

De Saussure also made careful measurements of any differences in how much of each gas was taken in or given out in the light as compared to the dark in green plants. Scientists were confused for a very long time by the fact that at all times, day and night, plants are respiring (taking in oxygen and giving out carbon dioxide), while they are photosynthesizing only during daylight hours (taking in carbon dioxide and giving out oxygen). Measuring plant respiration from green tissues in the light, then, has always been a challenge. Many plants seem not to give out carbon dioxide during the day from their green tissues because whatever amounts of the gas they produce in respiration during daylight hours is immediately used inside the plant in photosynthesis without being released into the surrounding air. Likewise, oxygen is being produced in photosynthesis during the day; at least some of this is used immediately in respiration within leaves. Separating respiration from photosynthesis by measuring the gases taken in and given out by leaves during the day is next to impossible, therefore, even with modern sophisticated instruments. In de Saussure's day, the task was impossibly bewildering to many people.

To make matters even more confusing, we now understand that, even when carbon dioxide *is* given off from green plant tissues during daylight hours, it may originate in so-called *photorespiration,* a process which has nothing to do with true respiration at all but, instead, with photosynthesis itself. Photorespiration provides no energy to the plant but causes some of the carbon dioxide being processed in photosynthesis to be released again before it can be used for producing carbohydrate. In effect, it is an inefficiency in the use of carbon dioxide in photosynthesis in many plants. Thus, some of the carbon dioxide released by photosynthesizing leaves during the day comes not from respiration at all but from photorespiration, a quite different matter.

For these sorts of reasons little progress in knowledge of respiration in plants took place in the first half of the nineteenth century because the exchange of gases associated with photosynthesis and the opposite exchange in respiration were confused with one another. In fact, *both* were referred to for a long time as respiration. In addition, photorespiration was not even suspected at that time so its contribution to the production of carbon dioxide by plants went completely unrecognized. Not until the latter half of the nineteenth century was a start made to resolving this confusion and true respiration (oxygen in; carbon dioxide out) separated in green plants from the opposite exchange of gases in photosynthesis (carbon dioxide in; oxygen out). By this period in history, the respiration of animals, by contrast, was much better understood in part because of the absence of complicating factors like photosynthesis.

Respiration, like photosynthesis, proved to be far more complex than anyone imagined before the facts of the process were known. There are approximately 50 steps if all aspects of respiration are taken into account. In an attempt to make matters simpler, most explanations of respiration are limited to how glucose, one of the main carbohydrates formed in photosynthesis, is broken down again to release energy. The series of steps leading from glucose to the final release of carbon dioxide, water, and energy is very similar in all types of organisms from the simplest bacteria to the most complex animals and plants.

However, we should remember that in many living things, not only carbohydrates, such as glucose, are burned in respiration but fats and proteins as well. This is particularly true of those animals where a highly active lifestyle places severe demands on the energy supply system. In such cases, carbohydrates, fats, and proteins are combusted.

In plants, proteins are not so routinely burned for their energy. Maybe this is because plants produce such large quantities of carbohydrate and, at the same time, have such a difficult time finding the nitrogen they need for protein formation in their surroundings (see chapter 4). Perhaps they do not lose their nitrogen during respiration as easily and quickly as animals do for this reason.

Proteins may not be used in respiration in plants but fats certainly are in particular cases. This is true especially in certain seeds (soybean, canola, and mustard, for example) where fats stored as droplets of oil are used up during seed germination. In fact, on a weight-to-weight basis, fats pack more of an energy punch than do carbohydrates. In most plants, however, the use of fats in respiration is not great. In the vast majority of plants carbohydrates seem to be by far the main source of material for combustion.

Why does respiration have to be so complex? After all, it is easy enough to release the great amount of energy contained in sugars. All you have to do is set fire to glucose or sucrose; they will both burn up quite readily releasing nearly all their energy as heat. If you throw a small pinch of sugar crystals (sucrose) onto an open fire, for example, a hot, bright flame will flare up at once.

Unfortunately, this method of releasing energy is not very useful to most living organisms. Occasionally, an organism may want to turn most of the energy released in respiration towards the formation of heat (see some examples later in this chapter). But conditions inside the cells of a plant, or any other living thing, are a far cry from uncontrolled conflagrations like those caused when sugar is thrown into an open flame. Inside most cells, the temperature must remain moderate otherwise the cell will die. Only a few organisms are specialized so that they can live in the extreme temperatures found in such environments as hot springs and deep-sea volcanic vents. As well, the main constituent in a cell is water. A great deal of heat would hardly be of use in such an environment. The main function of respiration, in fact, is to trap as much as possible of the energy given off in the burning of carbohydrates and fats in a form that will do the organism some good.

This is what the 50 or so steps in respiration achieve. Larger molecules, like carbohydrates and fats, are broken down to smaller molecules such as carbon dioxide and water. In the process, some of the energy released (around 40 percent) is retained in a form that can be used by the organism wherever an input of energy is required. In animals, these needs may include the demands of muscle activity, growth of the body and its organs, temperature regulation throughout the body, and the working of the brain. In plants, the greatest demand for energy comes from the areas where new growth is occurring (such as root and shoot tips) and where, at certain times, flowers, fruits and seeds are being manufactured. The rest of the energy released in respiration, a majority in fact (some 60 percent), is given off to the environment as heat in all organisms.

If we consider just the breakdown of glucose then as a representative example, the main pathway of respiration can be divided into two parts. In the first, glucose is *partly* broken down in a series of about nine steps. Some energy is released in this series, but not very much, and *no oxygen is needed.* Even so, this is an important set of steps which may date back many billions of years to a time when there was no oxygen in the atmosphere on our globe. The ability of organisms to release stored energy without the need for oxygen may have been very important at that time. Today, we see remnants of that more restricted lifestyle in the process of *fermentation,* more of which in a moment.

In the second part of respiration, the partial breakdown of glucose begun in the first stage is completed, all the way to carbon dioxide and water, a process that *requires* oxygen and releases a great deal of energy. It is this stage of respiration which provides to all organisms growing in air, not just plants but animals and microbes as well (so-called *aerobic* organisms), the supplies of energy they need to carry out their life processes. This energy currency is the same in all types of living organisms; it seems that a single high energy molecule (called adenosine triphosphate or ATP for short) has been adopted universally within the living world to deliver the energy released during respiration everywhere it is needed.

Some organisms live under conditions where oxygen is either not available to them or is severely restricted. This can be true of some green plants, under conditions of waterlogging of soil, for example, or in water where there has been a depletion of oxygen. Some seeds, when they are just beginning to germinate inside thick, hard seed coats, grow under conditions of restricted oxygen until the young seedlings burst out from the restricting seed coats and are exposed to the surrounding air. Under these *anaerobic* conditions, the breakdown of glucose follows the same path as in the first part of the aerobic process outlined above; the breakdown is only partial. But then, instead of being broken down further to carbon dioxide and water, as it is in the presence of oxygen, the partially broken down glucose is channeled into other products, usually ethanol.

Of course, we call this production of ethanol, fermentation, a process which has been used in wine and beer making for many centuries and is something which we associate particularly with yeasts. In fact, it was German scientists interested in alcohol production who first discovered this path of glucose breakdown in the first 40 years or so of the twentieth century. The process requires no oxygen but does produce bubbles of carbon dioxide (as wine- and beer-making afficionados well know) and releases some energy in the form of ATP which is useful to the organism. Heat is also released, again as wine and beer makers appreciate.

Although fermentation, or anaerobic respiration as it is also known, can be a significant energy-producing system in many microbes, including the yeasts, it is not nearly efficient enough to sustain the more vigorous life-styles of green plants; even less so of mobile organisms like animals. Here, the demands for energy are so much greater. Aerobic respiration is a necessity in organisms of these types.

Aerobic respiration releases nearly 20 times more energy than does its anaerobic counterpart. Thus, most green plants which find themselves in conditions where oxygen is restricted (as in waterlogged soil, for instance) cannot survive for very long on anaerobic respiration alone unless they are adapted for living in environments where oxygen is normally in short supply. Not only is there not enough energy being released to maintain most plants under such conditions but the ethanol accumulated in water-logged roots can be quite toxic and will kill the root after a short time if not removed. Even the most efficient yeasts specifically bred for use in the brewing and wine-making industries cannot survive more than about 23 percent alcohol in their surroundings. Most green plant tissues are killed by much lower alcohol concentrations than that.

Aerobic respiration, then, in plants as well as animals and most microbes is the principal way in which energy, in the universal form of ATP, is generated for everyday life processes. Respiration in plants, however, has some features that are not commonly found in other living things. These differences are exploited to their advantage by green plants.

In just about all organisms, including some plants, aerobic respiration is quickly poisoned by chemicals like cyanide and carbon monoxide. Both of these substances have been favorite agents of death in murder mystery novels down the years, and in real life, too! Curiously, though, many plants seem immune to the effects of these poisons. If plant tissues are exposed to cyanide, for instance, respiration gives the appearance of continuing almost completely unaffected for a period of time. This so-called *cyanide-resistant respiration* is found not just in many green plants but also in some fungi, algae, and bacteria as well as a few animals.

Cyanide and carbon monoxide both stop respiration by blocking the steps in which oxygen is used and which lead to energy release. In most living things that depend on aerobic respiration this blocking action quickly leads to death. Plants, too, have the steps leading to oxygen use and energy release blocked by these poisons. Why then do the plants not die right away as most other living things do?

The answer is that green plants have an alternative way of using oxygen and releasing energy in respiration. What is different is that this pathway does not lead to the release of much energy in a form that a plant finds useful for its normal life processes (i.e., ATP). Rather, most of the energy released in plant respiration in the presence of cyanide or carbon monoxide appears as heat.

This cyanide-resistant, heat production pathway can be beneficial to certain plants for particular reasons. For example, in the case of the arum lily, like skunk cabbage *(Symplocarpus foetidus)*, heat is used in a pollination strategy. When the arum lily flower is ready for pollination, its temperature can rise as high as 30 °C. The heat causes certain chemicals in the flower to evaporate; the scent released into the air attracts potential pollinators (see chapter 13).

Other plants also use the warmth created through the cyanide-resistant respiratory pathway to make themselves more desirable. For instance, certain arctic plants create cosy spots for insects by stepping up flower temperature by as much as 8 °C. In this way, the mountain avens *(Dryas octopetala)* attracts the nectar-eating mosquito which is also a pollinator. In a similar way, the arctic poppy *(Papaver radicatum)* makes itself attractively warm for pollinators. Many plants, such as crocuses, which bloom very early in the spring in cooler temperate climates, produce enough extra warmth to help melt snow in their immediate surroundings by raising their temperature by a few degrees.

So, the alternative pathway of respiration does have its practical uses. However, cyanide and carbon monoxide will kill plants eventually, just as they do other kinds of living things. Plants, like all other organisms, cannot live by heat energy alone. Heat is a useless form of energy for most life processes.

More controversial is the question of whether the alternate pathway of respiration serves any purpose under normal conditions. After all, many more plants have the pathway than use it for tasks such as generating heat to aid in pollination or to help melt snow. What do these plants use it for?

One clue may be that the pathway has been found to have its highest activity in plants that are rich in carbohydrates, for example after rapid photosynthesis. As a consequence of this rapid carbohydrate production, respiration, the breakdown of carbohydrate, is also greater. There is speculation that the speed of carbohydrate breakdown might overwhelm the normal route of energy release; some believe the alternate pathway may act as an overflow to relieve this pressure on normal processes. While having the appearance of being wasteful of energy accumulated at great expense in photosynthesis, it may be that plants are not able to control their release of energy in respiration in an any more efficient way than this. At least this release of surplus energy in the relatively useless form of heat would not harm the plant in any way but would, rather, act as a safety valve.

It would be wrong to leave the impression that respiration begins and ends with the release of energy, whether that be in a form which the plant can use to help drive its life functions or as heat. Mention was made earlier of the fact that respiration involves about 50 steps and that at each step another chemical compound is formed which can then do one of two things.

Each step in the respiratory chain can simply lead to the next, moving inexorably towards the consumption of oxygen and the release of carbon dioxide, water, and energy, as has been described in this chapter so far. Alternatively, each of the chemicals formed along the way provides an opportunity to the plant to siphon off some percentage of these products and use them for other things. After all, chemicals temporarily removed from the respiration chain leading to energy release could be returned to that chain later, just as we can take water from a river, use it for many purposes and then return it to the river to allow it to resume its journey to the ocean. These products from the respiration pathways are used in the formation of thousands of new compounds in plants, some of which will be discussed in later chapters.

For example, waxes on leaves, colored pigments like chlorophylls and carotenoids, terpenes such as rubber, and hormones like the gibberellins, are one huge grouping of compounds formed from the same source in the pathways of respiration. Fats and oils also arise from this same source and can be converted back into carbohydrates such as glucose, sucrose, starch, and cellulose. In other words, the route by which carbohydrates are broken down to carbon dioxide and water in respiration is not entirely a one-way process. This mobility can be important in sending valuable carbon wherever it is needed to build new plant structure.

Chemicals produced at other points in the respiratory pathway leading to carbon dioxide and water can be siphoned off to form the amino acids needed for protein formation, the building blocks of the genetic material deoxyribonucleic acid (DNA) and ribonucleic acid (RNA), and other important molecules such as the hormones auxins, ethylene, abscisic acid (ABA), as well as the alkaloids such as caffeine, nicotine, and many more.

Respiration, then, is much more than just a way to provide the energy necessary for the life of the plant. It is the source of many of a plant's building materials as well.

It would also be wrong to leave the impression that respiration proceeds in all plants and in all parts of any one plant inexorably, at the same speed, day and night. In fact, many factors influence the rate of respiration which can vary just as it does in ourselves. The more vigorously we exercise, for example, the higher our rate of respiration. During sleep our respiration slows down.

Of course, respiration depends on having something suitable to break down to carbon dioxide and water. Starved plants low in starch or stored sugars respire more slowly and speed up when supplied with carbohydrates just as we do. For this reason, the respiration of leaves is fastest just after the sun goes down when carbohydrate levels are high after the day's photosynthesis. As expected, respiration is often lowest just before the sun rises. Shaded leaves low on a plant often have a slower rate of respiration than leaves higher up the stem where they are exposed to higher light levels and where, therefore, more carbohydrate has been formed.

In extreme starvation conditions even proteins can be respired but this is a last resort in plants as it generally is in animals. People on hunger strikes tend to use up the carbohydrate and fat reserves in their bodies before any substantial breakdown of proteins. After all, the main occurrence of proteins in the animal body is in the form of muscle tissue and the thousands of enzymes needed to keep the machinery of the body working. In plants, maintaining muscle tissue is not the big issue it is in animals but plants do have enzymes as do all living things. Breaking down enzymes, nearly all of which are proteins, to provide energy makes little sense since without enzymes the energy would be useless anyway. But, as a last resort, some less essential proteins will be broken down to provide energy to those crucial, core life processes that the plant attempts to maintain and protect to the bitter end.

Changes in rate of respiration also occur during development of ripening fruits. In all fruits, the respiration rate is high when they are young while the fruit is growing rapidly. The rate then declines as the fruit matures. However, in many species, this gradual decline in respiration is reversed by a sharp increase at the time of full ripeness and flavor of the fruit. Common fruits that show this rise in respiration, called the *climacteric,* at the time of full ripeness, include apples, pears, peaches and nectarines. Conversely, citrus fruits like oranges, lemons, and grapefruits, as well as cherries, grapes, pineapples, and strawberries do not. Why some fruits have the climacteric and others do not is unclear.

Earlier, I introduced the process of fermentation and the fact that it occurs in conditions where oxygen is either not available at all or is in very short supply. This raises the question as to how sensitive plants are to the presence or absence of oxygen in their surroundings and how they ensure delivery of an adequate supply of oxygen to all their tissues.

Slight variations in oxygen concentration in normal air probably have no measurable effect on plants. There is also generally no problem in maintaining a constant, steady, high level of oxygen in most leaves, stems, and roots. But bulkier tissues, such as carrots, potato tubers, and other storage organs, tend to have a lower rate of respiration deep in their centers than closer to their surfaces. Yet, there is still enough oxygen available from the air at the center of a carrot or a potato for respiration to be aerobic, not anaerobic, even in these very dense tissues.

Air spaces are important in storage tissues and in organs which normally grow in conditions where oxygen is less freely available. In potato tubers, for example, about 1 percent of the tissue volume is occupied by air spaces; in roots from a wide range of species air spaces can be seen to occupy between 2 and 45 percent of the root volume. The higher values are common among plants growing in wetlands where wide air tubes or air spaces may extend all the way from the leaves down into the roots. Air taken in through stomata in leaves (see chapter 1) can then be moved efficiently down to roots which may be immersed in water where oxygen is depleted. Air spaces of any kind are important in the movement of gases around the plant to where they are needed or, as in the case of the carbon dioxide produced in respiration, from where they can be removed.

Some plants are better adapted than others to withstand oxygen depletion. For example, grasses and sedges have hollow stems and tend to withstand flooding well, therefore. Air can be supplied to the roots down the stem under flooding conditions in these plants. Among crop plants, only rice is known to tolerate oxygen depletion for long. This is partly due to the fact that rice seeds can germinate under conditions of low oxygen by relying on a very rapid and efficient fermentation for energy release. The young seedling can then take this energy and, in turn, channel it towards those tasks that are essential for rapid growth. In this way, the seedling grows swiftly above the surface of the water it is in and can then take in air directly and move it to submerged roots, directly. This is achieved due to the fact that rice has wide air tubes that extend from the leaves into the roots, which are then well supplied with oxygen even when they are submerged in water, as they are for long periods in paddies.

Even large plants like shrubs and trees have different tolerances to oxygen availability. In some tropical mangroves there are roots that grow straight upwards out of the water in which the trees live more or less continuously. Air is transferred down through these roots to the remaining submerged root tissues. Among the conifers, lodgepole pine *(Pinus contorta)* is more tolerant to flooding than is the Sitka spruce *(Picea sitchensis)*, for example. The main difference, again, is the greater ability of the pine to transport oxygen down into its roots.

Thus to summarize, there is much more to respiration than just a simple exchange of oxygen and carbon dioxide. Larger molecules, such as the carbohydrates, fats and, *in extremis,* proteins, are oxidized to carbon dioxide and water in the most complete form of respiration; that which depends on oxygen. When oxygen is unavailable or is restricted in supply, other endpoints than the formation of carbon dioxide and water may be found. What we call fermentation, the formation of alcohol, is one way many organisms, not just plants, use to generate the energy they need for growth. Fermentation, however, is much less efficient a route of energy release than aerobic respiration and is not sufficient to sustain organisms as large and vigorous as most green plants. Therefore, as an alternative, some plants have ways of moving air to areas where oxygen may be restricted. Thus, air tubes or spaces within tissues allow the transport of air from the aerial parts of a plant to its roots where less oxygen may be available from the surroundings. Cyanide-resistant respiration is another route of energy release common in the plant world although not unique to it. In this pathway, energy is released mainly in the form of heat which has limited value to a plant. Some plants use the warmth produced in this type of respiration for particular tasks but mostly, it is regarded as an overflow from regular respiration.

Once again, though, it must be stressed that respiration should not be viewed simply as a way to generate energy. Along the pathways of respiration, substances are formed which are the foundation of branch routes leading to thousands of other organic compounds. In many cases these substances are known to play crucial roles in the life of the plant; in thousands of other cases, roles are only now being understood; and in yet other examples, the reasons for their formation are quite mysterious. The roles of some of the members of this cornucopia of carbon compounds are discussed in later chapters.

4

Nitrogen, nitrogen, everywhere . . .

It is a curious fact that, apart from water, nitrogen is *the* key substance that limits where and how well plants grow in different areas of the world. Of course this means that the distribution and survival of animals is also controlled by the availability of nitrogen since animals are dependent on plants for food, directly or indirectly.

Why is nitrogen so crucial to the living world? What critical roles does it play in living systems?

The short answer to such questions is that two of the major components of living systems, proteins and nucleic acids, contain nitrogen. The machinery that builds, drives, and sustains all living systems is directed from nucleic acid blueprints, the genetic program in DNA and RNA, within each living cell. The machinery itself in each cell is made almost exclusively from protein, notably the enzymes which direct the thousands of chemical reactions carried out in living organisms.

Both nucleic acids and proteins contain nitrogen. The quantity of nitrogen available in usable form is a major determining factor, therefore, in how much nucleic acid and protein are made within an organism. Since there is much more protein in an organism than there is nucleic acid, it is the limit to protein production that is the most critical.

Nitrogen is enormously abundant. Nearly 80 percent of our atmosphere is made up of nitrogen gas and even that is only about 7 percent of the total nitrogen on earth. Nearly all the rest can be found in the rocks of our planet. Why, then, is nitrogen such a limiting factor in the world of living plants?

Nitrogen was given the name 'azote' by the French chemist Lavoisier, a word meaning 'without life'. By giving nitrogen such a name Lavoisier drew attention to the contrast between the two most abundant gases of the atmosphere, oxygen and nitrogen. He found oxygen to be very active and nitrogen boringly inactive (or inert) by comparison. This is one of the notable characteristics of nitrogen in its pure form as a gas. Only when nitrogen is combined with other elements, like hydrogen in ammonia or oxygen itself in nitrites and nitrates, does it become more reactive. Three chemical bonds hold the two atoms of nitrogen together in a molecule (N_2) of the pure element and are so strong that very large amounts of energy have to be used to break them and release the two trapped nitrogen atoms so that they can react with other substances. These bonds *can* be broken, as we shall discover later, but not by any animal or plant, directly.

In addition to the inert nitrogen gas in the atmosphere there is an enormous quantity of the element trapped in the rocks of the earth as mineral salts. This reservoir of nitrogen is continually being set free into the soil, waterways, and atmosphere by the slow processes of weathering of rock by wind and water, and by the more explosive but haphazard process of volcanic action which may blast dust and gases containing nitrogen over a wide area, even the entire globe on occasion. These natural processes may release some nitrogen from rock but vastly greater amounts remain locked away.

Estimates have been made that of the total amount of the earth's nitrogen (a huge 57×10^{18} kg), 93 percent is locked away in the rocks, while nearly all of the remaining 7 percent (3.8×10^{18} kg) is in the atmosphere as inert nitrogen gas. Only a tiny fraction of 1 percent (1.5×10^{15} kg) is present in soils and waterways where it is accessible and in a form which can be used by living things, mainly nitrates, the most highly oxidized form of nitrogen.

So, the reason why nitrogen is not accessible to most organisms is that most of it is locked away in rock deep in the earth and most of the rest is in the inert form of nitrogen gas in the atmosphere.

An additional problem is that even when nitrogen is available to the living world in the form of nitrates, for example, it is *still* not accessible to many kinds of living things. Animals, including humans, are unable to use either nitrate or ammonia for their nitrogen needs. In animals, nitrogen *must* be in an organic form before it can be used; that is, the element has to be combined in biological molecules along with carbon (see chapter 2 for a definition of organic molecules) before it can be used by animals. Ready-made protein is the main form of organic nitrogen available to us, as to other animals. An infant born at, say, 3.5 kg will become an adult of 60 kg or more; the difference includes about 11 kg of protein. To make this protein we, and other animals, must take in nitrogen either as protein or some other form of organic nitrogen in our food.

Not only does the average adult human body have in it some 11 kg of protein at all times but the protein it does have is also constantly being broken down again within the body and rebuilt. We can refer to this as protein 'turnover', the constant renewal of the machinery of life. It should be emphasized that this is a process common to all living things, not just animals like ourselves. Unfortunately, during this breakdown and rebuilding process some nitrogen is lost from the body. Thus, even the mature adult whose weight remains constant still requires a regular supply of dietary protein to replace the continuing losses from this turnover.

In different tissues and for different proteins, the renewal process may occur at intervals ranging from every few minutes to days or even months. In other words, some proteins turnover very quickly, others more slowly. The total turnover of protein in the average adult is about 300 grams (g) per day of which about 40 g is lost from the body, mainly in urine, faeces, sweat, and menstrual fluid, and the shedding of skin, nails, and hair. Diet, therefore, should contain at least 10 percent protein to replace the lost nitrogen. And not just any old replacement protein will do. Most importantly, the proteins we eat must contain an adequate supply of, what are called, the *essential amino acids,* an added complication for animals like ourselves.

The molecules we call proteins are built by joining together smaller units, called *amino acids,* into long chains. Twenty different kinds of amino acids are needed to build all the kinds of proteins found in living systems. Plants can manufacture all 20 for themselves but animals, only some. The rest must come from the proteins animals take in through their diets. Generally, it is recognized that ten of the amino acids we, as humans, need we cannot make for ourselves. Gaining these, so-called, 'essential' amino acids[1] is not a problem for those of us with rich diets. The crops and livestock that supply the bulk of our energy, vitamins, and minerals provide our protein as well. In rich countries and in those that specialize in raising livestock, people obtain about one-third of their energy from animal products and nearly three-quarters of their protein.

In poorer countries, animal products typically supply less than one-tenth of the daily energy, and perhaps only one-eighth of the protein. Poor people often do not have access to adequate supplies of meat and sustain themselves largely on a diet of plants and plant products. World-wide, then, and especially in poorer countries, plants provide the human race directly with about 80 percent of its daily energy and two-thirds of its protein, as well as useful amounts of minerals and vitamins. Thus, there is a widespread dependence on plants to provide not just organic nitrogen in our diets but nitrogen in the right form as well; the essential amino acids.

Of all the crops we cultivate, the cereals are, generally, the best foods. They provide energy in the form of starch and sometimes fat as well as several vitamins and minerals. They also contain between 8 and 12 percent protein. This amount of protein would satisfy the average dietary need of the human race, which as already pointed out is around 10 percent of total daily dietary intake, except for one small but crucial point.

Cereals are usually lacking in the essential amino acid, lysine. In addition, some varieties have less than a 10 percent overall protein content. This may not create a critical problem for adults in underfed countries of the world although mature men and women in these regions would undoubtedly benefit from extra protein. The main problem is with babies and growing children who need proportionately more protein than adults. The disease kwashiorkor, which we have all no doubt seen examples of through television news reports and documentaries from areas of famine and malnutrition, is a direct result of protein deficiency. Who could forget the images of the hugely swollen abdomens of underfed children suffering from this terrible disease. Fortunately, in most regions of the world, a remedy to the lack of lysine in cereal grains is to hand.

In addition to the cereal grains, in all parts of the world where plants are the main source of food, crops such as the pulses are cultivated to supplement the diet. Beans, peas, lentils, and chickpeas are pulse crops of particular value. The protein of these pulses (or legumes, if you prefer) perfectly complements that of cereals. The protein from pulse seeds is particularly rich in the lysine that cereals lack.

So, pulses and cereals eaten together provide an excellent protein balance. All the world's popular cuisines include mixed cereal–pulse dishes: chapatis and dhal in India; frijoles and tortillas in Mexico; and beans on toast in many Western countries. In addition, soybean has become one of the most important of all cash crops. Soybeans are in great demand because they can be processed in so many different ways to make foods. Textured soy protein can be flavored to make anything from dog food to hamburgers, and both oil and 'milk' are also extracted from the soybean. In Japan, miso, a soup traditionally eaten at breakfast time, is made from fermented soybean paste. Tofu, made from soybean milk, is also eaten in soups with rice, meat, and vegetable dishes. Soy sauce is made from fermented soybeans.

So, once again, we have been brought back to plants as the source of something crucial to the needs of other organisms in the biosphere, in this case nitrogen in the form of protein and amino acids. We have seen this essentiality before in the case of energy supply in photosynthesis (chapter 2) and we shall encounter it again from a different point of view in later chapters dealing with perfumes (chapter 13) and especially medicinal plants (chapter 16). Plant protein can satisfy human dietary needs directly, in the form of cereals and pulses, or indirectly, via the animal protein we eat. Ultimately, all of this derives from plants, anyway.

The question, then, eventually comes down to one of how do plants gain their nitrogen?

For carnivorous plants, the answer to this question is a straightforward one: turn the tables on animals and eat *them* instead of them eating you! Insects attracted to the pitcher plant by the sickly sweet smell of its nectar slide helplessly down the slippery surface of the bell-shaped pitcher into the acid bath at the bottom where they drown. As the flesh of the animal decomposes the nitrogen released is absorbed and used by the plant. Pitcher plants, including the Venus flytrap *(Dionaea muscipula)*, sundews *(Drosera* species) and bladderworts *(Utricularia* species), all live in bogs where the ground is so acidic and waterlogged that bacteria cannot finish the job of breaking down dead plant matter into nitrogen-rich soil; a far less nutritious peat is formed which cannot supply the plant inhabitants of the bog with the nitrogen they need. As a result, the carnivorous plants must draw their nitrogen from animal prey. All have their individual ways of attracting and, then, engulfing their victims; only the goal remains the same in all cases, the endless quest for nitrogen.

For the remaining members of the plant kingdom, the answer to the question of how plants gain their nitrogen requires that we first know something about what is called the 'nitrogen cycle'.

In brief, the nitrogen cycle starts with the breakdown in the soil of dead plants and animals by microbes, primarily the bacteria and fungi. These breakdown processes release nitrogen from the proteins of the dead organism into the soil in the form of nitrate and ammonia. Some of this nitrogen is returned directly to living plants via their roots where it is, once again, converted into amino acids and, then, proteins and other nitrogen-containing substances needed by the plant.

But not all of the soil nitrogen is returned to plants. Some is converted by soil microbes into nitrogen gas which then drifts off into the inert nitrogen pool of the air and is, essentially, then lost to living systems. Obviously, if this loss to the atmosphere were to continue, within a short time, all the nitrogen available to living systems in the soil would end up in the atmosphere and be lost. The fact that this does not happen is because some living organisms have found a marvellous, natural way of taking nitrogen directly from the air, breaking the very strong bonds between the nitrogen atoms, and then converting the nitrogen to a form usable by living things. We have also found a way to accomplish the same thing artificially in the manufacture of fertilizers. This process of *nitrogen fixation* I shall return to shortly. But first, the release of nitrogen into the soil.

When plants and animals die, their remains are broken down by microbes, mainly bacteria and fungi, in the soil. The nitrogen in these remains is 'mineralized', a term meaning that it is converted mainly into ammonia. In the case of animal flesh, this process, called putrefaction, may produce not just ammonia but also other, more foul-smelling products. The breakdown of plant material is usually less obnoxious, may take several years to accomplish, and gives rise to the humus content of soils.

In most soils, the ammonia produced by mineralization does not stay around for very long which is fortunate since ammonia is extremely toxic both to animals and to plants if present in the environment in too high a concentration for too long. The, so-called, 'nitrifying bacteria' in soils quickly convert ammonia to nitrite and then finally, nitrate. They do this to extract energy from the ammonia for their own use.

The advantage of nitrate to a plant is that it is not toxic. The disadvantage is that nitrate is very soluble in water and is quickly leached from soil by rain or irrigation. In this way, nitrate is rapidly carried deep down into the soil, where it can no longer be reached by plant roots. Alternatively, it is carried in waterways to the oceans. In agricultural practice, this lost nitrate is replaced either by adding fertilizer to arable land or through crop rotation with legumes which, as we shall see shortly, return nitrogen to depleted soils.

Unfortunately, the over use of nitrogen and phosphorus fertilizers in modern agricultural practice has led to the leaching of excess nitrate and phosphate into lakes, rivers, and streams in many parts of the world. In waterways, high concentrations of nitrate and phosphate cause, in turn, an explosive growth of algae and aquatic plants, a process called *eutrophication.* When these great masses of vegetation finally die they are decomposed by equally excessive numbers of microbes. The microbes use up much of the dissolved oxygen that fish and other aquatic animals need and they, then, also die, often quite suddenly and in great numbers, leaving the water devoid of life. The 'killing' of natural waterways by eutrophication is one of the greatest threats to the environment in any region where the profligate use of fertilizers prevails. Nitrogen may be essential to plant life but an excess of it can be highly detrimental.

Nitrate in moderate quantity, then, is harmless to plants but not, unfortunately, to young children. The very nitrate leached out of soil into ground water or waterways may end up in drinking water. Bacteria in our intestines can change this nitrate into nitrite which bonds very strongly to hemoglobin in the blood. The strongest bonds are formed with the kind of blood found in very young children (the fetal blood) which persists in infants for some time after birth. Hemoglobin with nitrite bound no longer functions and produces a form of anemia which can cause serious health problems in an infant.

Thus, although mineralization leading to the release of nitrate into the environment is a necessary part of the cycling of nitrogen in nature, the overapplication or overproduction of nitrate in soil through excessive fertilizer addition can be a serious matter in more ways than one.

Under some soil conditions, certain bacteria in the soil can cause the release of nitrogen from nitrates in the form of nitrogen gas. When soils are waterlogged, for example, they become short of oxygen. Some soil bacteria in flooded fields and pastures respond to this lack of oxygen for respiration by breaking down nitrate. Of course, they are seeking the oxygen that is found in nitrate, not the nitrogen. They take the oxygen for themselves and release the nitrogen as a by-product into the atmosphere. By doing this, however, the nitrogen is lost to other living organisms.

So, either by leaching of nitrates from soil or by the breakdown of nitrates to form nitrogen gas, there is a continual drain of usable forms of nitrogen away from the biosphere. If this nitrogen were not replaced soil would eventually be totally depleted of this essential element. Fortunately, various ways exist to replenish soil nitrogen, both naturally (nitrogen fixation) and artificially (the Haber industrial nitrogen fixation process).

Most of the nitrogen fixed industrially these days is used in the production of fertilizers. One recent, tragic exception to this tendency occurred in April, 1995, in Oklahoma City in the USA where a Federal Government building was destroyed by a bomb explosion with great loss of life. The bomb used in this attack seems to have been made from a mixture of common nitrogen fertilizer and petroleum products which, as illustrated in this incident, can explode with enormous force under the right conditions.

In fact, a major impetus leading to the setting up of the first industrial plants to produce ammonia from nitrogen gas was the need for munitions in World War I. Hydrogen has to be added to nitrogen to convert it to ammonia. Fritz Haber, a German chemist, in 1909 found a way to do this efficiently. The first commercial plant to manufacture ammonia using the Haber process was set up in Germany in 1913. At high temperature, under considerable pressure and in the presence of a catalyst, nitrogen and hydrogen combine to yield ammonia. The hydrogen can be provided either from natural gas, coal gas, or petroleum, all readily available ingredients. Today, millions of tonnes of ammonia are produced by the Haber process for peaceful use in agriculture.

The main product of the Haber process, ammonia, can be applied directly to soil, unchanged, or converted before use to other products like nitrate or urea. Without the addition of nitrogen in this way as fertilizer many arable lands around the world would be far less productive than they now are.

But what about the natural environment? How is extra nitrogen added to soil in nature? As we saw earlier, the mineralization of animal and plant material in the soil leads to the recycling of nitrogen in a form that can be reabsorbed by living plants. These plants may be, in turn, eaten by animals to satisfy their own nitrogen needs. But this recycling process does not lead to the addition of *extra* nitrogen to the living environment. The weathering of rocks adds some, slowly, and, these days, air pollution, from automobile and industrial emissions, can be washed down by rain and add yet more nitrogen to soil. But these sources are minor on a global scale. Fortunately, nature has devised an alternative upon which the natural world depends and which is hugely exploited in agriculture as well. We call it *nitrogen fixation.*

Earlier, it was pointed out that seeds of the pulses can be used directly in diets to supplement the protein provided by cereals. Legumes in general, both wild and cultivated, including the pulses, also can add nitrogen to soil both as the plants grow, and when they die and decay. The beneficial effects on the soil of legumes and the importance of green manuring was realized by the ancient Chinese, Greeks, and Romans. The use of legumes in crop rotation was well established, therefore, long before the reason why they were beneficial was discovered.

The French scientist Jean-Baptiste Boussingault (1802–87) must be given credit for providing the first firm evidence for the fixation by legumes of nitrogen directly from the air. Boussingault found that he could grow crops like clover and peas in soil which had no available nitrogen. The plants flourished and somehow gained an ample supply of nitrogen for their needs. In contrast, when he grew cereals like wheat and oats in the same type of soil they did not flourish and showed no ability to gain any nitrogen. Boussingault concluded that the nitrogen gained by the legumes

> *is derived from the atmosphere; but I do not pretend to say in what precise manner the assimilation takes place.*

For several decades after this discovery by Boussingault, work on the role of nitrogen from the atmosphere in plant nutrition continued, and more and more scientists became convinced that legumes were capable of fixing free nitrogen from the air.

The problem remained a baffling one, however. Then, in the late nineteenth century, two German investigators, Hermann Hellriegel and Hermann Wilfarth, announced their discovery of the role of certain bacteria found in the swollen nodules that are so characteristic of legume roots.

Hellriegel and Wilfarth grew peas in sterilized soil (soil in which all living organisms, including bacteria, had been killed by heating). Pea plants grown in such soil did not do well and had no nodules on their roots. Other pea plants were grown in similar sterile soil but this time the two investigators added to the soil a water solution, a leachate, taken from unsterilized soil in which other pea plants had been grown without any trouble. The pea plants exposed to leachate also grew well and developed nodules on their roots. When Hellriegel and Wilfarth examined the pea root nodules carefully, they found them to be teeming with bacteria whereas roots grown in sterile soil conditions had no nodules and, therefore, no bacteria. They came to the inescapable conclusion that it was the bacteria in the nodules of legume roots which somehow took nitrogen from the air and fixed it into a form useful to their pea plants. How the bacteria achieved such a thing remained a mystery for many years.

We now understand that two major groups of microbes are capable of fixing nitrogen from the air. First are the microbes that live free in the soil or in waterways. Among these are the blue–green algae (now called cyanobacteria since they have more in common with bacteria than with the algae) which live on wet surfaces including the surface of soil. Other cyanobacteria live in loose association with a variety of plant species. For example one type lives inside the leaves of the tiny fern called *Azolla* which floats in the water of rice paddy fields. This fern with its associated cyanobacterium provides most of the nitrogen needed by the rice during its life-cycle. Other cyanobacteria live on or in the roots of several evergreen plants as well as inside ferns (other than *Azolla*), lichens and liverworts. The cycad, *Macrozamia,* found in eucalyptus woodlands in Australia, fixes nitrogen through an association with cyanobacteria lodged in its above-ground roots.

Certain loose associations between free-living soil microbes, other than cyanobacteria, and plant roots occur in many areas of the globe. Nitrogen-fixing bacteria have been found growing in large concentrations on or near the root surfaces of crop plants such as maize, wheat, sorghum, rice, sugarcane, and other grasses. There seems no doubt that the plants benefit from these associations by having enhanced growth. The microbes gain by being supplied with sugars for their energy needs produced in photosynthesis by the plants they are living alongside in the soil. In return, the nearby plants have a guaranteed source of a scarce resource.

It is thought that as much as 100 kg of nitrogen per hectare per year may be fixed in arable lands where these kinds of loose soil associations occur. Particularly useful relationships of this type exist between certain microbes and the roots of some tropical forage grasses like *Panicum* and *Digitaria* as well as crops grown in the tropics such as maize, wheat, rye, sorghum, and millet. In some of these cases, the microbes can be found thriving inside as well as outside the plant roots.

The second group of nitrogen-fixing microbes are those with a special, intimate relationship with a host plant through the formation of nodules within which nitrogen fixation can take place. Plants of the Leguminosae (Fabaceae) family, such as fodder legumes like alfalfa (lucerne), clover, and the vetches, as well as the big-seeded pulse crops such as peas, beans, lentils, chickpeas, and peanuts, fall into this group. Shrubs such as gorse and broom, and trees, including the extremely important acacias of Africa and Australia, all harbor bacteria of the *Rhizobium* type in their roots. These were, indeed, the microbes in the leachates used by Hellriegel and Wilfarth in their early investigations of nitrogen fixation in the root nodules of peas.

We now know that the rhizobial type of bacteria (of which there are many) live free in the soil. However, unlike other soil microbes capable of fixing nitrogen directly, such as the cyanobacteria, rhizobia do not fix nitrogen from the air in their free-living state. They take in their nitrogen in the same way as any other soil organism, directly from the soil itself, as nitrate. Rhizobia exist quite happily in this way until a legume is planted into the ground close to where they are living.

As the legume seedlings develop their roots start to secrete substances into the soil which attract the rhizobia nearby. The bacteria eventually enter the roots and stimulate the formation of swellings, the nodules, inside which the microbes multiply. At the same time, the bacteria take on different shapes to such a degree that they no longer look much like the soil rhizobia from which they came. For this reason, in the roots they are called 'bacteroids' and these now have the ability to fix nitrogen from the air.

The bacteroids accumulate at the center of each root nodule and are surrounded by tissue that often takes on a noticeable pink coloration which turns out to be a form of hemoglobin. It closely resembles the hemoglobin found in animals like ourselves but since it is found in legumes it has been given the name *leghemoglobin.* Why do these particular plants need a substance that is normally associated with the transport of oxygen and carbon dioxide in animal respiration?

The answer is that the fixation of nitrogen by bacteroids is a process which is stopped dead by the kinds of oxygen concentrations found in normal air. On the other hand, the bacteroids in root nodules are in as much need as any other living things for oxygen for their own respiration. How are these two conflicting requirements balanced against one another? How can a supply to bacteroids of just enough oxygen for respiration be maintained without destroying their ability to fix nitrogen from the air?

In our blood, oxygen is carried by hemoglobin to the tissues of the body where it is needed. When it reaches its destination, the hemoglobin gives up its oxygen and then returns to the lungs for more. In the nodules of legume roots, leghemoglobin does not circulate but is contained in a tissue which completely envelops the region of the nodule where the bacteroids are living. As air percolates into nodules from the soil, the leghemoglobin intercepts the oxygen but allows the nitrogen in the air to pass freely through to the bacteroid colonies at the center of the nodule. The leghemoglobin then delivers oxygen to the bacteroids but at a controlled, reduced rate; just fast enough to satisfy the respiratory needs of the microbes but not in such a flood as to inhibit the nitrogen fixation going on inside the bacteroids.

The result of all this is that the bacteroids are able to fix nitrogen from the air. Of course, they also need an ample supply of energy to use to make the nitrogen gas reactive. For this they rely on the host plant which transports sugars from its leaves down to its roots. With these sugars the bacteroids satisfy their own energy needs and, in return, produce more ammonia by nitrogen fixation than they need. The ammonia is made available to the host plant which converts it into a usable organic nitrogen form in the nodule before transporting it to other parts of the plant.

As might be expected, then, the conditions that favor rapid photosynthesis in the host legume plant also favor nitrogen fixation by bacteroids in root nodules. This is because the energy needed for fixation is provided by the 'burning' during respiration within the bacteroids of sugars produced in photosynthesis in the leaves of the plant host and transported down to root nodules. So, good moisture conditions, moderate to high temperatures and bright sunlight all favor nitrogen fixation in those plants with the ability to perform this miracle.

It is, arguably, not too strong to call this process miraculous. The nitrogen fixation carried out by certain microbes at normal temperatures and pressures in and around plant roots we can match only through the use of enormous amounts of energy in the industrial Haber process; temperatures of 300–400 °C and pressures greater than 350 atmospheres.

The rate of nitrogen fixation is at its maximum during the early afternoon when the movement of sugars from photosynthesizing leaves to other parts of the plant is occurring most rapidly. Early afternoon is also the time of day when transpiration often occurs at its fastest. The rapid removal of the products of nitrogen fixation from the roots in the fast-moving water stream (see chapter 1) up to the transpiring leaves helps stimulate the fixation of more nitrogen by root nodules.

The first product of nitrogen fixation is ammonia which, as we have noted previously, is very toxic to plants. For this reason, ammonia is converted to amino acids and other nitrogenous substances in roots before being transported to the rest of the plant. Nitrate taken in from the soil, on the other hand, is not toxic to plants. Nitrate, therefore, may or may not be converted to ammonia and then amino acids in roots before being moved elsewhere within the plant system. Nitrate can just as well be transported in the water stream from the roots to the leaves before being converted to ammonia and amino acids.

Measurements of grasses and legumes indicate that there is extensive daily recirculation of nitrogen from roots to leaves and back again. This constant movement is likely to be important in directing nitrogen to the areas of the plant where it is needed most and to prevent any particular part of the plant from becoming short of nitrogen. These movements include not just the redistribution of new nitrogen coming in as nitrate through roots from the soil or as ammonia from root nodules but also the turnover of protein nitrogen within the plant itself. As mentioned earlier in the chapter, protein molecules in all living systems are constantly being broken down into their smaller amino acid units and then rebuilt. The amino acid products of this turnover can be moved to other parts of the plant if there is a need for protein building blocks somewhere else.

This amino acid release process is especially important at certain times of year. Nitrogen gained by the plant during the leafy growth phase of its life cycle can be mobilized and moved from roots, stems, and leaves into flowers and then fruits, seeds, and storage organs when the need arises. Nitrogen is moved wholesale out of dying leaves in the fall in temperate climates, and stored in the roots and stems of the plant for use the following spring. In some annuals, the movement of nitrogen out of leaves can be very extensive. Wheat leaves, for example, lose as much as 85 percent of their nitrogen to flowers and seeds before they die.

One of the enduring enigmas of biology is why the tremendously important ability to fix nitrogen from the air possessed by certain microbes has not been transferred to plants in the course of the hundreds of millions of years of evolution. It is estimated that over 100 million tonnes of nitrogen are made available to the living world each year by fixation in free-living microbes in the soil, and by natural and agricultural plant–microbe associations such as those found in the legumes. Nitrogen is required in quite large amounts by plants, and limits plant growth in both natural plant communities and arable lands throughout the world. Yet only microbes can do it. Why?

One answer to this question may lie in the fact that nitrogen fixation is very costly in terms of the amount of energy which must be supplied to the bacteroids in root nodules. About eight times as much energy is needed to fix every molecule of nitrogen as is required to fix each molecule of carbon dioxide in photosynthesis. Thus, most plants do not associate with microbes to fix nitrogen the way legumes do. The great majority of green plants take their nitrogen directly from the soil, mainly as nitrate. Even legumes and other plants that do fix nitrogen stop doing so when supplied with adequate amounts of nitrate. One of the features of legume crops, for example, is that their productivity cannot easily be increased by adding fertilizer to them during the growing season. This is because the nitrogen supplied as fertilizer does not significantly add to the amount of growth supported by nitrogen fixation. Instead, the plant shuts down fixation until it uses up the ready made source of nitrogen added as fertilizer; nitrogen fixation is restarted once fertilizer in the soil is depleted. It is an either/or proposition, not an additive one.

Not that the conversion of nitrate to ammonia necessarily has any lower energy requirement than the fixation of nitrogen from the air to form ammonia. It may simply be that the processing of nitrate occurs within the plant itself whereas nitrogen fixation is taking place through an association with microbes. Maybe nitrate processing is, therefore, easier to accomplish. There is no fully satisfactory answer to this question.

What is clear is that nitrogen fixation by soil-borne microbes is essential to the cycling of nitrogen out of the atmosphere and into the environment occupied by living organisms. Green plants provide a crucial link in this cycling through their direct participation in host–microbe associations as well as by absorbing forms of nitrogen from the environment (nitrates and ammonia most notably) and converting them to the kinds of organic forms (amino acids, and proteins) that animals and other non-green organisms can use.

Nitrogen may be everywhere in great abundance but without the intervention of green organisms, including plants, it would remain azote, as Lavoisier so aptly put it.

Notes

[1] Lysine, leucine, isoleucine, phenylalanine, tyrosine, methionine, cysteine, threonine, tryptophan, and valine.

5

Nutrition for the healthy lifestyle

If we think of 'good nutrition' at all it is likely because of the constant bombardment these days in newspapers and magazines, and on radio and television from health workers of one sort or another, exhorting us to maintain a balance in what we eat, as one important aspect of a well regulated lifestyle. From all this, we have a vague notion that to eat immoderately and injudiciously is, somehow, harmful.

Usually, however, it is in the long-term that any possible harm to our health exists. There is rarely any immediate consequence, other than a possible mildly annoying case of indigestion, arising from the fact that we indulge ourselves from time to time by overeating or eating foods that we have been told by experts are likely to harm us. Hence, we continue to stray from the nutritional straight and narrow while being aware of that mean voice in our minds telling us that, likely as not, we shall be sorry for this in the future, sometime.

We are also generally aware that plants become sick, just as animals do, when not supplied with the nutritional requirements they need for their continued good health. For animals, these requirements are elaborate and include the balanced provision of complex molecules in their diets, such as carbohydrates, proteins, and fats, as well as vitamins and certain minerals.

In the case of plants good nutrition is rather different from that for animals. As we have seen already in chapters 2, 3, and 4, green plants are capable of producing their own sugars and other carbohydrates as well as fats and proteins from simple molecules, like carbon dioxide, water, oxygen, and nitrogen. Plants are capable of manufacturing their own complex molecular machinery of life internally rather than through dietary supplements.

Just because they have these capabilities within themselves, however, does not mean that plants can exist with no other input from the environment than light, water, carbon dioxide (for photosynthesis), oxygen (for respiration), and nitrogen. In common with animals, plants also need certain minerals for their continued healthy growth. Some of these are essential to both plants and animals in greater or lesser amounts; others are essential either to animals or plants but not both; and some are of probable, but at present uncertain, value to either.

We are generally aware that we eat some plant materials for their mineral content, like bananas for their potassium and spinach for its iron, minerals that are essential both to animals and to plants. Sodium is essential to animals but not to plants, which can be grown to maturity without it. Molybdenum is essential to plants but is toxic to animals when more than a trace is present in food. Aluminum, although found in both plants and animals, is not considered to be essential to either. In fact, aluminum is toxic to plants and may also be detrimental to animals. For example, its concentration in the brain tissues of sufferers from Alzheimer's disease has led to speculation about its possible role there.

We are also generally aware that mineral deficiencies in plants are caused by the lack of a particular nutrient (such as potassium, nitrogen, or iron) in the soil in which the plant is growing. We have surely all had the feeling of helplessness as our favorite house plant or garden shrub slowly shriveled before our eyes despite our best efforts to nurture it. Of course, whether those yellowing or curling leaves are a sign of a fatal disease or a lack of some nutrient in the soil in which the plant is growing is often very difficult to discover. And the symptoms are just as likely to be due to the fact that we forgot to water the plant for a week or two! All we can do is add some general fertilizer, containing a wide variety of minerals, along with water and place our faith in the ability of most plants to stand astonishing amounts of abuse and yet survive.

We are also familiar with the practice of throwing far too much fertilizer (usually a mixture of nitrogen, phosphorus, and potassium) on our lawns to provide an unnaturally lush, green sward. Farmers throughout the world enhance their crop yields through the repeated application of nitrogen, phosphorus, and 'potash' (potassium) to arable land deficient in these nutrients.

The symptoms of mineral deficiency in humans are sometimes quite easy to see and simple to reverse. We all know what to do if we are anemic; take iron for the blood. A lack of the correct balance of calcium, phosphorus, and vitamin D can lead to rickets in children or osteomalacia in adults, but a deficiency is simple to remedy by eating foods rich in these nutrients. We use iodized salt in food preparation and eat iodine-rich foods such as sea-fish and the seaweed kelp to help avoid iodine deficiency that can lead to the condition called goiter which is associated with the function of the thyroid gland. In some parts of the world salt (sodium chloride) is still a form of currency and is highly prized because of its dietary significance. Herbivorous wild and domesticated animals seek out, or must be provided with, 'salt licks' to satisfy their craving for this essential nutrient which cannot be obtained from plants. These and many other examples point to our general recognition of the need animals like ourselves have for certain minerals.

Similarly with plants, certain abnormal conditions can be recognized as being linked to a deficiency in particular minerals. For example, 'heart-rot' of sugarbeet (boron deficiency), 'dieback' of citrus trees (copper deficiency), 'whip-tail' of cauliflower (molybdenum deficiency), and 'little leaf' of apple (zinc deficiency) all have recognized sets of symptoms caused by shortage of specific minerals and can be cured by supplying the appropriate one.

Ideas about nutrition in general can be traced back to Aristotle who believed that both plants and animals took in food in various combinations of the four elements: earth, air, fire, and water. Since plants did not seem to have the kinds of organs animals had, stomach and intestine for instance, for changing the food taken in to fit their purposes, Aristotle decided that they must absorb it in a form perfectly suited to their needs. To Aristotle, further evidence for this was the fact that plants produced no waste as excrement, unlike animals; plants simply sucked up perfect 'nutrient fluid' from the earth.

Such a view of plant nutrition prevailed for many centuries, as did many of Aristotle's conclusions, right or wrong. His view of plant nutrition began to crumble during the sixteenth and seventeenth centuries as the scientific style of investigation took hold. Until that time, carrying out experiments was believed to be beneath the dignity of the great men of the day. In any case, questioning the opinions of such revered sages as Aristotle was regarded as heresy.

A major turning point was the, now, classic experiment of van Helmont in the seventeenth century in which he concluded, erroneously as it turned out, that the whole substance of a plant is formed from water alone (see chapter 2). Prior to van Helmont, the unchallenged view since the time of Aristotle was that the bulk of a plant came directly and solely from the substance of the soil. Van Helmont showed, however, that over a 5-year period, his willow tree increased in bulk by nearly 75 kg yet the soil in which the tree was growing decreased in bulk by only a few grams. He concluded from this that, therefore, the increase in weight of the willow must have come from the water added to the soil during the 5 years of growth and could not possibly have come directly from the soil itself. Van Helmont, of course, knew nothing about the contribution of air to the nutrition of plants.

Van Helmont was right in saying that the bulk of a plant did not come directly from the soil but wrong in concluding that it, therefore, came solely from water. Despite being wrong in his final conclusion, his well conducted experiment raised many questions about how plants do obtain their food.

One of the first to challenge van Helmont's final conclusion was John Woodward, a Professor of Medicine living in London at the end of the seventeenth century. Woodward questioned whether the nutrition of plants came from water itself, as van Helmont contended, or from what the water contained. For example, when he grew spearmint, potatoes, and vetch in water from various sources (springs, rivers, rain, sewer effluent, and after purification by distillation), Woodward found that plants grew better in water containing impurities than in pure, distilled water. In his own words, most of the water that enters a plant:

> *passes through the pores of them* [the stomata of leaves] *and exhales up into the atmosphere; that a great part of the terrestrial matter* [solids, including minerals, from the soil] *mixt with the water passes up into the plant along with it; and that the plant is more or less augmented in proportion as the water contains a greater or smaller quantity of that matter.*

Thus, Woodward came to the conclusion that water alone was not sufficient to sustain a plant; the 'terrestrial matter' the water had dissolved in it was also of importance to the health of plants.

And so the issue remained for about the next 100 years, until the early part of the nineteenth century. Then, the Swiss scientist, Theodore de Saussure, became one of the first to attempt a more systematic investigation of plant mineral nutrition than any of his predecessors. For example he grew plants of lady's thumb *(Polygonum persicaria)* in water which had in it just one type of mineral (potassium alone or iron alone, etc.) and discovered that not all the minerals were absorbed by the plants in equal amounts. By making this simple observation, de Saussure discovered that plants were capable of selecting what they took in from the soil; they could discriminate between one mineral and another.

De Saussure also insisted that particular minerals among those absorbed were absolutely essential to the growth of his plants while others were not. Thus he also discovered that plants were capable not only of discriminating between minerals, but also selected in favor of minerals they could not function without.

These careful studies about the importance to plants of mineral nutrients in the soil created a great deal of debate among scientists in the first half of the nineteenth century and led to much additional evidence in support of de Saussure's views. One of the most significant further conclusions came from a German investigator, Carl Sprengel, who wrote that a soil may be favorable in almost all respects:

> *yet may often be unproductive because it is deficient in one single element that is necessary as a food for plants.*

Here was a clear statement for the first time of the need plants have for a balanced array of nutrients, the lack of any one of which could hinder their growth. And here also was a hint of, what we would call now, agricultural science, the systematic investigation of plants and their requirements for optimal growth.

Investigations on plant nutrition in the mid-nineteenth century reached a high point through the efforts of the Frenchman, Jean-Baptiste Boussingault, who is the person generally credited with having laid the foundation of what we now call agricultural science. He was not content with studying just the mineral composition of plants as his predecessors had done. Boussingault stressed the importance of the balance between the amounts of each mineral absorbed by crop plants and the amounts they extracted from the soil in which they were growing. He also laid the foundation of the need for fertilizer to maintain the balance of nutrients in soil by replacing the different amounts of each mineral lost to plants during a growing season.

Despite his painstaking work, Boussingault's ideas about the balanced mineral requirements of plants were greeted either with scepticism, indifference or, in the case of the foremost chemist of the day, Justus von Liebig, with outright ridicule. Liebig, never one to be shy about putting forward his own views no matter whether well-founded or otherwise, went so far as to express the old, discredited views of Aristotle, a full half century after de Saussure had shown them to be untrue, namely that:

> *All substances in solution in a soil are absorbed by the roots of plants, exactly as a sponge inbibes a liquid and all that it contains, without selection.*

By this time, both de Saussure and Boussingault had shown that such a simplistic view of plant nutrition was not believable. Plants do not just take in minerals as a sponge would, without discrimination.

Fortunately, Liebig did better, later, in saying that:

> *Plants live upon carbonic acid* [carbon dioxide], *ammonia (or nitric acid), water, phosphoric acid, sulfuric acid, silicic acid, lime, magnesia, potash and iron.*

Thus, in his usual assertive, authoritative fashion, Liebig put forward the view that finally won acceptance for the 'mineral theory of fertilizers', something de Saussure and Boussingault had been unable to do with the same finality. Liebig considered that soil contributed soluble *inorganic* (certain mineral) constituents to plants, not *organic* material (the 'nutrient fluids' of Aristotle). In this, as in many other matters despite his occasional lapses, Liebig was correct.

The question that could not be answered at the time was whether Liebig's list of inorganics (ammonia, nitric acid, etc.) was complete. Were these the only minerals needed by plants?

Questions about which minerals from soils are absolutely essential to the good health of all plants continue to arise even today, although we do now have a much more complete understanding of the matter than did scientists in Liebig's day and before. One major problem in determining whether a particular mineral is essential or not to plants was solved with the development of, what we would call today, *hydroponics.*

In the latter half of the nineteenth century, Julius von Sachs, a German botanist, discovered a way to grow plants entirely in the absence of soil, putting to rest once and for all any idea that plants needed solid organic material from the soil. He prepared solutions of mineral nutrients and grew plants directly in these liquids, without soil. Growing plants in water solution culture (hydroponically), as it became known, has been a favorite technique in plant nutrition ever since. It seems to be especially favored at present by those wishing to conceal their caches of marijuana plants which grow very well under artificial light, in solution culture, in remote or concealed locations!

The advantage of the solution culture method is that the complexities of soil can be eliminated by growing plants without it. Soils of all kinds are very complex and it is impossible to know in every detail what they contain. For example, soils have in them most of the known chemical elements. In addition, while we know now that plants can be picky about how much of a particular mineral they take in from the soil, they are not so good at excluding any one of them completely. Tiny amounts of all minerals in a soil are likely to find their way into the plant. Over 60 of the known chemical elements have been detected in plants.

The question then becomes, which of these minerals are essential and which are non-essential? To answer that question, von Sachs' solution culture method was particularly valuable.

The mineral content of solutions, unlike soil, can be very closely, although not completely, controlled. Starting with very pure water, known amounts of particular minerals can be added to the water, or, equally importantly, left out. What von Sachs did was exactly that. After finding out which combination of minerals seemed to maintain plant growth in his hydroponic solutions, he then left out each mineral in turn and observed the effects on his plants. In this way he, and and those who followed him, discovered that in addition to carbon, hydrogen, and oxygen, plants seemed also to require, what became known as, the seven 'ash elements'; phosphorus, potassium, nitrogen, sulfur, calcium, iron, and magnesium.

Any analysis will show that most of the solid bulk of a green plant is composed of just a small number of chemical elements. The 'big four' (carbon, hydrogen, oxygen and nitrogen) make up 95 percent of a plant's dry weight. If plant material is incinerated at very high temperature, the big four are burned off (except for some of the nitrogen), leaving behind an ash which contains nothing much more than the ash elements listed above. For several decades into the twentieth century, these seven were thought to be the only minerals essential to plants. Now we know better.

What early investigators in the field of plant nutrition could not know is that some minerals are required by plants in the minutest imaginable quantities. For instance, it is estimated that a plant needs 60 million times less molybdenum for its growth than it does hydrogen, yet both are essential. Such minute amounts could not be detected by the crude methods of measurement at the time the ash elements were being discovered.

Now it is clear that, in addition to the seven ash elements, all of which are required in relatively high amount except for iron, there is another group of minerals, called the *micronutrients,* which are needed in only the tiniest amounts by green plants. This group includes molybdenum, copper, zinc, manganese, boron, chlorine, and nickel. Whether the list is complete is impossible to determine; maybe not. It is possible that more will be discovered to be essential perhaps in even smaller amounts than in the case of molybdenum.

Generally, plants respond to an inadequate supply of one of the essential minerals by showing recognizable deficiency symptoms, at least to the practiced, professional eye. Most non-experts would have a hard time distinguishing, say, the symptoms of nitrogen deficiency from those for magnesium, I suspect. Deficiency symptoms include such things as stunted growth of roots, stems, or leaves, the yellowing of leaves, and the browning of various parts of the plant. But then certain plant diseases can cause similar distortions. Knowing which are a sign of mineral deficiency and which of disease takes practice and experience. The appearance of deficiency symptoms do, however, help professional agriculturalists, horticulturalists, and foresters decide which fertilizers to add to their crops, and when.

For example, magnesium deficiency shows up first in the older leaves of a plant which begin to turn yellow when this nutrient is in short supply. Magnesium is a part of the chlorophyll molecule; a shortage of the mineral means that chlorophyll cannot be formed.

Magnesium can be moved around inside a plant very easily, it is very soluble and mobile. Thus, when magnesium is in short supply, a plant will destroy the chlorophyll in its older leaves and move the released nutrient to younger, more vigorous leaves developing at the tips of its branches. Hence, the older leaves, having lost their chlorophyll and, hence, their magnesium, turn yellow while the younger leaves, having gained the mineral, continue to form green chlorophyll. Of course, if no more magnesium becomes available to these plants, through the addition of a fertilizer containing it for example, then the younger leaves eventually turn yellow, too.

The appearance of deficiency symptoms in older parts of a plant before younger ones generally indicates that the mineral in short supply is very mobile within the plant and can be easily and quickly moved around. Plants seem to prefer to move what essential minerals they can from older to younger tissues if the supply of these nutrients is insufficient for all their needs. In addition to magnesium, this is the case for nitrogen, phosphorus, potassium, and chlorine.

What about the other way round, the appearance of deficiency symptoms first in younger plant parts before older ones? Symptoms of this sort occur because some minerals, once used, cannot be released again and moved elsewhere within the plant.

For example, once calcium has been used within a plant it seems very difficult to release it again and move it somewhere else. In part, at least, this may be because much of the calcium is built right into the structure of the plant itself. In animals, like ourselves, large amounts of calcium are fixed in the bones. Plants do not have skeletons in the same way but the 'glue' that sticks together the individual cells of a plant contains calcium. We know this glue as the pectin gel that can be boiled out of fruits used to make jams and jellies. Thus, just as in the case of bone, once this glue has been formed and used, the calcium it contains is not likely to be set free again, otherwise the whole plant would become 'unglued', and this simply is not allowed to happen.

Therefore, a plant lacking calcium does not have the option, in contrast to magnesium, of releasing the mineral from older parts and moving it to the new. Thus, new areas of the plant show calcium deficiency symptoms (twisted and deformed stems, roots, and leaves) before older tissues.

Boron and iron, like calcium, are also difficult to move around but the remaining essential minerals cannot be categorized quite so easily. Their deficiency symptoms are not identified so directly with either older or younger tissues, rather they can occur in any or all parts of the plant.

Deficiency does not always mean that the level of a mineral is low in the soil around the plant either. Deficiency in some cases depends very much on conditions in the soil in which the plant is growing. For example, most soils of the world are rich in iron yet plants growing in many of these soils show iron deficiency, the main symptom of which is the yellowing (or chlorosis) of leaves.

Volcanic soils are rich in iron (many of them are reddish-brown in color, an indication of this) and it might be expected, therefore, that plants grown in soils of volcanic origin would not show iron deficiency symptoms, but they do. In the Hawaiian Islands for example, which are entirely volcanic in origin, the pineapple crop is sprayed with solutions containing iron several times in a growing season to prevent leaves from becoming chlorotic. The reason – Hawaiian soils also have high levels of manganese. Iron is not taken in by plants very well when high levels of manganese are also present. Hence, leaf chlorosis is common in these islands.

The name 'limestone chlorosis' has been linked to the fact that plants growing on alkaline soils, high in lime, fail to direct iron in sufficiently high quantity to where it is needed in the growing plant. To some extent, this effect can be linked to whether the soil around the plant is acidic or alkaline. Iron is much more easily taken up by plants growing on acid than on alkaline soils.

Large applications of phosphate fertilizer can also cause iron chlorosis even on soils rich in iron. Somehow, iron compounds in the soil are not compatible with phosphates in fertilizer mixtures. The iron and phosphate react together in the soil to form substances that plant roots cannot absorb.

Thus, at least for iron, soil acidity or alkalinity, the abundance of other minerals like manganese in the same soil, or which fertilizer is added to a soil, are all factors that can affect its absorption by plants. These factors may not be related in any way to the amount of iron present in a particular soil. And similar factors apply to all other minerals that plants require from the soil. As is so often the case in nature, 'balance' and 'moderation' are two watch-words for a healthy existence, even among plants.

Finally, the crucial part plants play in making minerals available to other organisms needs to be recognized. In this, plants play an irreplaceable role.

In chapter 1, we saw that the roots of plants invade almost every particle of soil within a large volume, often to considerable depths. One of their purposes in doing so is to seek out the water the plant requires. At the same time as it is sucking in water from the soil the plant also takes in whatever is dissolved in the water. This includes the minerals which are, thus, brought up, perhaps from considerable depths below the soil surface, into the plant roots and then to all parts of the plant. There, after serving an array of functions in the life of the plant, some of the minerals in the leaves find their way back into the surface layers of the soil during autumn leaf fall or through year-round leaf and twig fall in cases where the discarding of plant parts is not linked to the seasons. In other cases, the death of the whole plant can lead to the wholesale delivery of minerals into the upper layers of the soil.

In any event, minerals are in this way 'mined' from deep in the soil and eventually deposited at the surface where, when plant litter is mulched, the minerals released can be used either by other plants or different organisms which are growing in the same soil but which cannot penetrate very deeply into the earth. Also, by eating the leaves and other parts of plants, herbivores and microbes satisfy their own mineral needs after which they, too, return material to the soil surface as waste or when they die.

In this function of cycling minerals out of the soil to where they are available to all other living organisms we can say that, as in photosynthesis, plants are indispensable to the stabilization and future of life at the earth's surface. Without this constant retrieval of essential minerals from the depths of the soil, leaching by rain would lead eventually to a depletion of nutrients at the surface and the severe curtailment, if not elimination, of life there.

The efficiency with which plants carry out this 'mining' also depends very much on soil acidity. In the first place, the rate of growth of roots is favored by mild acidic soil conditions. We also now understand that most roots of deciduous plants are surrounded by and live in very close contact with particular fungi which aid greatly the roots' ability to take in minerals and, probably, also water. These fungi act as a kind of extra root surface since they deliver to the root on which they are growing the minerals that they have taken in from the soil but which are surplus to their own needs. The growth of these fungi is also favored by acidic soil conditions.

Finally, soil acidity favors the weathering of rocks and the release from rock particles into soil water minerals like potassium, magnesium, calcium, and manganese. In addition, carbonates, sulfates, and phosphates are more soluble in soil solution under acidic conditions. All of these factors contribute to increasing the availability of minerals in the region of the soil occupied by plant roots.

We began by considering that good nutrition involves the provision of certain foods and dietary supplements to all organisms but especially animals, including ourselves, and plants. For animals these requirements are quite elaborate; they overlap with those of plants at least in the need for certain minerals. The amounts of minerals taken in are related closely to the needs of an animal or a plant. For example, the ratio of the amount of calcium in the soil to that found in plants and in humans is 1:8:40; for phosphorus, 1:140:200; and for sulfur, 1:30:130. These ratios confirm what de Saussure discovered long ago; each kind of living thing takes in the minerals that are essential to it in the amounts needed to maintain optimum growth.

Apart from satisfying their own needs for certain essential minerals, plants also play a crucial role in the essential cycling of these nutrients from deep in the soil to surface layers of our planet where the majority of organisms of all kinds live.

Good nutrition is something we all need, plants included. As in many other respects, we can see that in addition to supplying minerals for their own survival, plants also ensure that other living organisms obtain the minerals needed to maintain good health.

6

Transport of delights

Green plants flourish if supplied with air, light, certain minerals, and water. Mostly, they take in the water they need through their roots and transport it all the way to their leaves, then out into the surrounding air. Along with this upward flow come the essential minerals from the soil needed for continued good health. The gases they need (carbon dioxide and oxygen) are brought in from the atmosphere via leaf stomata. Through photosynthesis, mainly in their leaves, plants manufacture their own food, principally sugars, using carbon dioxide and water as raw materials. In turn, the sugars formed are used to supply the entire plant with energy and the chemical building blocks needed to produce other essential molecules needed for growth, such as fats, proteins, and nucleic acids.

Growth can take place in all tissues of a plant throughout its lifetime. Roots continually spread within the soil; main stems and side branches continue to grow in length and width; new leaves are produced either seasonally or all year round depending on the type of plant; and flowers, fruits, seeds and, in some cases, storage organs, like bulbs, corms, and tubers, are formed periodically. Food is needed in all parts of plants to sustain these activities.

Water and minerals are, thus, not the only substances that must be moved around in the plant in an orderly and timely way. Certainly, water and minerals from the soil must reach the highest branches of the tallest trees and we have seen how that is accomplished in the upward water stream (see chapters 1 and 5). But what about the food produced primarily in the leaves by photosynthesis? Only green parts of the plant can photosynthesize; many other parts are not green. They, too, must have food to provide their energy and other needs. How does a plant move food from where it is produced to other places where it is required? Before answering that question we must understand the nature and scope of the problem.

For example, what about the fast-growing tips of branches where leaf buds are found? These buds contain many tiny, young leaves which are too immature to carry out much, if any, photosynthesis, yet are growing rapidly. For this, they need lots of energy and a steady supply of food from older, more mature leaves. The trunks and stems of trees and many shrubs are not green yet contain substantial amounts of living tissues which must somehow be provided with nourishment. At certain times of year, plants form flowers and then seeds neither of which may be green, certainly not when they are mature. The food needed to produce and then sustain them must be brought in from elsewhere in the plant. Some plants produce substantial food storage organs like bulbs, corms, and tubers, to which, as they swell, food must be delivered.

These various demands for food can arise not only at widely separated locations in a plant but also at different times during a growing season. In temperate climates early in spring, for example, the main need of a growing plant may be to move food from mature leaves that are capable of photosynthesis or from storage organs to young, developing leaves at stem tips, and to rapidly growing roots. Later, as the plant begins reproductive growth, flowers may have to be nurtured followed by fruits and seeds. Later still, if overwintering storage organs are formed they, too, must be supplied with significant amounts of food. All of these organs can be at different locations on the plant. Whatever supply system the plant has, then, must be versatile enough to deliver the food required not just in a downward direction to roots and storage organs but upwards to developing stem tips and across to developing flowers, fruits, and seeds.

Then there are the leaves themselves which, when young, import food to fuel their rapid expansion but which, as they mature and start full production of sugars in photosynthesis, become food exporters. Here, the delivery system must be capable of transporting food *into* the leaf at one time and *out of* it sometime later.

The question then becomes, where in a plant is there a distribution system with the flexibility and versatility needed to deliver food to all parts of the plant to support all its functions, and what form does it take?

The discovery of human blood circulation by William Harvey in the first half of the seventeenth century was a great stimulus to the study of circulatory systems in general, including plants. Yet, it is only in recent decades of the twentieth century that we have begun to understand the details of how foods and other kinds of substances are moved around the plant, a process that has come to be known as *translocation.* Why was the problem so difficult to solve? Partly it was because translocation occurs within a system of tubes that are microscopic in size and located deep inside a plant where they are difficult to observe.

In animals, the arteries and veins of the blood system are frequently visible even to the naked eye. In plants this is not so. The hearts of many animals with their large chambers and associated plumbing are relatively easy to find and study. Even here, a major stumbling block at first to a deeper understanding of how the system worked was the microscopic capillaries in the tissues of the body that connect the arterial system with the veins. To early observers of blood systems, there seemed no obvious connection between arteries and veins and, therefore, no clue as to how blood could circulate between them. Only when the role of the tiny blood vessels we call capillaries was understood to be to form the bridge between arteries and veins could the entire circulatory system be visualized.

In plants, there is really *only* a capillary system, which is, because of its microscopic structure, difficult to find and even more difficult to study closely. There is nothing equivalent to the animal arteries and veins in plants although we use the word 'vein' to describe the appearance of the circulatory system within a leaf. The word is used loosely, however; the veins of a plant leaf are not at all like those of an animal blood system.

As we have discovered in earlier chapters, from the time of Aristotle until about the mid-seventeenth century, it was thought that plants obtained their nutrients, ready made, directly from the soil. About then or a little later, individuals like van Helmont and de Saussure made observations which contributed to a drastic change of opinion on this point. To their names must be added one more, the Italian scientist Marcello Malpighi, who, around 1675, made some telling observations about the translocation and role of sap in plants.

Malpighi *girdled* a tree by removing a complete ring of bark from around its trunk, without, however, damaging the underlying wood in any way. He noticed that over the following weeks the bark above the girdle became swollen and immediately below, thinner.

Malpighi also showed that 'ascending, or raw, sap', as he called it, from the soil was changed in the leaves, but only in sunlight, into an 'elaborated sap', which was then carried down to other parts of the plant. He also found that girdling blocked the movement of elaborated sap; the swelling of tissues immediately above the girdle resulted from the accumulation of this rich, downwardly moving sap. Malpighi also discovered that this type of swelling did not occur to nearly the same extent during winter months when the trees he had girdled had lost their leaves.

Malpighi also noticed that, although the downward movement of elaborated sap was blocked by girdling, water moving up from roots continued to reach the leaves. Unfortunately, he did not appreciate the significance of this discovery. Despite all the evidence they gathered, Malpighi, and others who followed him in carrying out similar girdling experiments, failed to see that the movement of water *up* a plant occurred in *different* tissues from those used to move elaborated sap *down* from leaves. Well into the eighteenth century it was still believed that the ascending and descending streams of fluids in plants traveled in the same channels. Even Hales, who provided so much of the early information on water movement in plants (see chapter 1), said, in his book *Vegetable Staticks,* half a century after Malpighi and after many girdling experiments of his own:

> *Upon the whole, I think we have, from these experiments and observations, sufficient ground to believe, that there is no circulation of sap in vegetables; notwithstanding many ingenious persons have been induced to think there was, from several curious observations and experiments, which evidently prove, that sap does in some measure recede from the top towards the lower parts of plants, whence they were with good probability of reason induced to think that the sap circulated.*

Despite these misunderstandings about movements of fluids up and down a plant, by the end of the eighteenth century it was generally understood that sap rising from roots was carried to the leaves of a tree or shrub in the vessels of the wood. There, in the presence of light and air, photosynthesis occurred and the sap was elaborated, made rich in sugars. It was then transported to other parts of the plant, providing nourishment for growth. What was not appreciated at that time was how it managed to get there.

During the third decade of the nineteenth century, a new conducting tube system was discovered in the bark of plants which was quite different and distinct from that for transporting water through the wood of the same plants. The tubes used for water transport in wood are simple open channels much like the pipes which carry water to and within our homes; they are dead tissue. In contrast, the tubes discovered in bark were living cells.

Soon, it was realized that these tubes in the bark were important in the translocation of Malpighi's elaborated sap. For example, if the tubes were cut they leaked sap rich in sugars. The amounts of sugars leaked were greater or lesser depending on whether the leaves were carrying out photosynthesis. In bright light conditions, photosynthesis was vigorous and the sap in the tubes in the bark, rich in sugars; in poor light or complete darkness, the content of sugars in the sap was much lower.

From many observations like these, the conclusion was at last drawn that the downward movement of sugars from leaves was taking place in these channels within the bark, a conclusion that we now accept and which has been confirmed many times since the mid-nineteenth century. We now understand that these channels, called *sieve tubes*, occur in most green plants, whether the plants have a recognizable bark or not. We now know, too, that wherever they are found they perform the same function of carrying predominantly the products of photosynthesis, but some other substances as well, from places where they have been formed or stored, such as leaves, to where they are needed. Organs like leaves which export the food needed by the rest of the plant have come to be known as 'sources'; 'sinks' are those areas of the plant which have to be supplied with food.

Once the significance of sieve tubes was understood, attention could be directed to answering questions such as, how quickly can the movement of sap take place in these tubes? After all, if they are living tissue plugged full of living matter, how can they act as pipes for the efficient movement of food around a plant? As already pointed out, we know that food has to be moved in all directions in a plant; from leaves to roots, from leaves to shoot tips, from leaves to flowers and fruits, from storage organs to all parts of a plant, from old leaves to young, and so on. And several of these movements may have to go on at the same time. How is it possible to have sap moving in sieve tubes in opposite directions at the same time?

Questions such as these about the speed and direction of movement of material in sieve tubes have occupied plant biologists for many years. Understanding how sieve tubes work has very practical value since we now know that not just food can be moved in these channels. For example, there is much current interest in the application of chemicals, like herbicides or substances that promote growth or fertilizers, directly onto leaves rather than through the soil. Many of these substances must then be transported from the leaves where they are applied to other locations in the plant where they have their effect. The sieve tubes play a role in this.

The spread within a plant of many of the diseases caused by fungi, bacteria, and viruses often involves the circulatory system. For instance, certain fungi that infect roots have their effect by producing toxins which are spread up the stem to the leaves via the water stream. Some viruses are introduced directly into the sieve tubes in the leaves or within stems by feeding insects (such as aphids, leaf hoppers and white flies) and then moved along with translocating food.

How sieve tubes act in translocation is important, therefore, not just for understanding how plants move their own products but for many other reasons besides.

The main challenge in answering questions about the performance of sieve tubes was how to observe them in a working, functioning state when they are microscopic in size, often deeply hidden inside the plant and surrounded by other kinds of tissues not in any way involved in translocation.

From the early days of investigation it was quickly learned that sieve tubes are very delicate. They are easily damaged if disturbed in any way and stop functioning immediately. This is frustrating for the investigator but may make sense to the plant. Damage to its translocation system may be a signal to the plant that it is under attack by disease organisms or by a predator such as an insect. The immediate response of the plant to this perceived danger is to stop the flow of the damaged part of the translocation system perhaps in an attempt to prevent the introduction of infection or to protect the rest of the plant from further injury.

Whatever the reason, the delicacy of sieve tubes means that they cannot easily be removed from plants to see how they work. They have to be examined wherever they are located. Ingenious ways had to be found to gain access to the translocation system without causing it to stop functioning.

One of the most effective ways of following the movement of food around a plant in sieve tubes is to make radioactive the sugars being formed during photosynthesis. This can be done quite routinely as long as great care is taken in handling the radioactivity!

Sugars produced during photosynthesis are formed from carbon dioxide and water (see chapter 2). We have had available to us radioisotopes of carbon for many decades. To produce artificially radioactive carbon dioxide from this carbon is not very difficult. In the hands of experts, radioactive carbon dioxide can be supplied to a plant and used by it in photosynthesis to produce radioactive sugars in leaves. Some of these sugars will then be moved out of leaves in the sieve tubes and find their way to other parts of a plant. The speed and direction of their movement around the plant can be followed by tracking the radioactivity.

Mainly through the use of radioactive sources, like carbon dioxide, we have learned that translocation of food in sieve tubes normally occurs at a speed of between 50 and 150 centimeters (cm) per hour. It is difficult to visualize how fast this is but by way of an analogy, a speed of 100 cm per hour is equivalent to the motion of the tip of a 16 cm-long minute hand on a clock. By watching a clock hand of this length carefully and patiently we would have no trouble seeing a rate of movement like this with the naked eye.

Put another way, if a sugar molecule were to be translocated 100 m down one of the tallest trees from a leaf to the roots at a velocity of 100 cm per hour the time taken would be 100 hours or just over 4 days.

It is probably not necessary for a plant to move food over quite such great distances. For instance, the lowest leaves on a tree are closer to the roots than those at the top of the tree and are more likely to be the sources from which the roots are supplied with the food they need. Thus, the distance of translocation may be much less than 100 m in the case of a tall tree. We must remember, however, that translocation is occurring in tubes which are microscopic in width, much narrower than the finest strand of hair, and are filled with living material rather than being open pipes.

One possible explanation of the rapidity of movement seems to be that very high pressures are maintained in sieve tubes. The highly concentrated sugar solutions in the tubes act strongly to attract water into the translocation system by osmosis (see chapter 1). The accumulation of water in this way helps maintain a pressure within sieve tubes of between 20 and 30 atmospheres. This very high pressure is used to force fluids through sieve tubes in whichever direction food is required.

The pressure is, of course, highest in the leaves where the sugars produced in photosynthesis are 'loaded' from the green areas where they are being formed into the sieve tubes located in the veins of the leaf. The sugar solution is then forced out of the source leaf under pressure towards the various sinks where it is needed. As the sugar solution is withdrawn at the sinks, the pressure in the sieve tubes drops.

Thus, food simply moves to wherever it is needed in the plant, from areas of high pressure to anywhere else in the plant where food is being withdrawn from the translocation system and where, therefore, the pressure is lower. This could be to the roots, or where flowers and fruits are being formed at the tips of branches, or where storage organs are being produced in preparation for winter. Flow of food in the translocation tubes changes direction simply depending on where the sinks for the food carried by those particular tubes happen to be.

From chemical analyses of sieve tube contents we have learned that 90 percent or more of the material translocated in plants is carbohydrate. Sugars are the most frequent of the carbohydrates translocated and are often at concentrations as high as a thin syrup (maybe as high as 30 percent) in sieve tube fluid, especially if, say, a leaf vein has just been loaded with sugars following a period of rapid photosynthesis on a sunny summer day. Further than that, nearly all the sugars being translocated in most plants are in the form of sucrose (table sugar).

This is not to say that other, less abundant compounds found in sieve tubes are unimportant. Some minerals, including nitrogen, usually accompany the sugars and can be just as important as the carbohydrates for the growth activities going on at a sink.

In a sense, then, translocation fluid is a balanced mixture which provides good nutrition to the non-green parts of plants.

Earlier, I said that the flow of food in sieve tubes changes direction simply in response to the location of the sinks, whether they be further up or down the plant from a particular source. But this is not quite the case. It has long been known, for example, that the leaves lower down on a plant tend to transport relatively more food downwards to roots than upwards to young leaves or to fruits and seeds; upper leaves tend to translocate preferentially upward to the youngest branches, developing fruits and seeds. This makes sense.

An important question, however, is how does the plant divide up its food (*partition* it) among all the competing sinks it develops during its lifetime? Somehow, delivery of adequate volumes of food have to be maintained to all competing sinks to allow the balanced growth of the entire plant. How is this achieved?

This is a question of great interest and practical importance to us as well as to plants themselves. The yields of agricultural crops depend on how much food produced in leaves is translocated to and then stored in those tissues of the plant which will be harvested at the end of the season, whether they be the seeds, fruits, roots, or other storage organs. Understanding what controls the partitioning of food within plants therefore is highly important knowledge to breeders of new crop varieties. If the partitioning process could be understood in enough detail then maybe it could also be manipulated through breeding to produce higher yields.

It is known that if potato tubers are removed early in their development before they have expanded much, photosynthesis in the leaves of the potato plant drops sharply. In other words, remove the storage organ sink, in this case tubers, and the supply of food is reduced accordingly. How might this sink demand regulate photosynthesis in the source leaves?

The simplest explanation would be that when demand is low in the rest of the plant, the back up of sugars in a leaf acts as a signal which causes photosynthesis in that same leaf to slow down. Without doubt, a regulation of the supply of sugars takes place, whether it be by this simple mechanism or some other more complex process. In recent years, evidence has been increasing which suggests that a special form of sugar, fructose, is used as a signal in leaves to regulate photosynthesis. The more sucrose is present in a leaf, the more fructose is also produced. As the concentration of the fructose increases, so photosynthesis slows down. So, if we could understand how this particular control mechanism works in crop plants we might be able to change it to prevent the action fructose has in slowing down photosynthesis thereby maintaining a high production of sugars and increasing crop yield.

So, we are beginning to have some idea of how food production in a leaf is regulated. What is not so clear is how the distribution of that food is controlled once it leaves the leaf in the sieve tubes. We have been able to improve the harvest yields of many species of crops in the last few years and can claim, therefore, that it is possible to breed for a more favorable partitioning of available food towards areas of the plant which will be harvested thereby improving the yield of a crop. Such changes have been accomplished in the case of oats, barley, wheat, cotton, soybean, and peanuts, to name just some examples.

What has not been so successful is breeding for increased photosynthesis in leaves so that the total amount of food available for translocation out of leaves is greater. We still have little idea how to do this.

Thus, plants do have a circulatory system for moving water, food, and other substances around. The system is complex and controlled in such a way as to deliver balanced mixtures of nutrients to all parts of the plant. It is efficient, fast, and responsive to the changing needs of a plant as it goes through its life cycle and is capable of satisfying the nutritional demands within even the largest plants.

7

Growth: the long and the short of it

The story is told of a traveler during the nineteenth century who came one night to an inn on the Yangtze River in China. Before dawn he was awakened by squeaking and screaming sounds in the grove of giant bamboo *(Dendrocalamus giganteus)* surrounding the inn. Alarmed by the noise, the traveler roused his companion who explained that the unusual sounds were simply the result of friction caused by the rubbing action of growing, young bamboo shoots as they forced their way out from between the tough protective sheaths enclosing them.

This story is a vivid illustration of the fact that growth in plants can be swift and vigorous although not usually as dramatic as in the case of the giant bamboo. The fastest growing trees are the eucalypts, one type of which, found in New Guinea, has been known to add nearly 8 m to its height in a year. However, even these 'sprinters' are eclipsed by giant bamboo which can grow over 1 m in a day and achieve 30 m in height in under 3 months.

At the other extreme is the example of a Sitka spruce found at the tree limit in the Arctic which had one of the slowest growth rates on record. From measurements of the annual growth rings in the trunk this specimen was estimated to be about 100 years old yet reached only 28 cm in height.

The growth of which some plants are capable in a lifetime is startling. One of the largest giant redwood trees found had a wood volume of more than 1500 m^3 and weighed over 1000 tonnes. Since the seed of the giant redwood weighs less than 0.005 g, the weight increase over the lifetime of this specimen was more than 250 billion times. In addition, these very large trees are capable of living for upwards of 4000 years illustrating that plants often combine in their structure tissues of great antiquity with others that are still youthfully producing new leaves, shoots, roots, fruits, and seeds.

Plant growth, then, is much more than just increase in size. As suggested in the preceding paragraphs, throughout their lifetime, whether that be 4000 years or only one as in the case of annuals, plants can produce enormous bulk at great speed. The example of the Sitka spruce in the Arctic, on the other hand, illustrates the fact that growth is not a fixed quantity but can vary (see also chapter 17). Low growth rate was undoubtedly related to the harsh climate.

But as they grow, plants also develop a variety of different organs, such as leaves, flowers, fruits, and seeds. In animals, organs develop very early in life and become at once an integral part of the whole organism without which it cannot function. In plants, organs are somehow programmed to be created and discarded repeatedly and more or less indefinitely as long as the plant lives. Many organs (leaves and flowers, for example) often come and go seasonally from the plant structure, disappearing at certain times of year and reappearing sometime later.

All living things, plants included, are programmed to take on particular shapes and growth forms. But in the case of plants, the internal genetic program which controls this is often remarkably flexible. One illustration would be the well-known ability of plants to produce roots under special circumstances, a power which gardeners make common use of when multiplying their plants by cuttings. Roots are not normally formed directly on stems and even less often by leaves. Yet a severed willow branch, when stuck in the ground will often 'strike' producing roots directly from the base of the cut end of the branch. Fence posts made from the tropical gombo-limbo tree *(Bursera simaruba)* have been known within a short time of being set in the ground to form roots and leaves and in a few years grow into lines of trees along the edges of fields. Leaves of begonia which happen to come in contact with the soil for a long enough time may make roots directly from the ribs of the leaf.

All of these activities of plants, whether straightforward slow or rapid growth in height, the orderly, seasonal growth of distinct organs, such as leaves, flowers, and fruits, or the seemingly unplanned formation of roots directly from cut stems and leaves, must be under some form of control. As has been said already, all organisms are programmed. Partly this programming is genetic. The fact that petunias look alike but different from oak trees is due, in large measure, to the differences in the genetic make-up of the two kinds of plants.

But in addition to genetic control of shape and form there must be another kind of control which is responsive to what is going on around an organism at any moment in time. For instance, in the case of a plant which normally flowers early in a growing season in a temperate part of the world, cooler than normal weather may cause a delay in flowering. In other words, such a plant is not genetically programmed so rigidly that flowering invariably occurs at precisely the same time, to the day, each year. Variations in temperature and moisture, for example, may force plants to slow down or speed up their seasonal programs controlling their growth and development. Flexibility is built into the programming to allow the plant to respond to variations in its environment. The contrast between giant bamboo in a warm climate with the Sitka spruce in the cold Arctic is one dramatic illustration of the variability possible.

The kind of flexibility illustrated in the case of flowering in the previous paragraph must be required for other growth responses as well. Somewhere, orders have not only to be issued for such things as more roots, shoots or leaves to appear, and fruits and leaves to fall in season, but also for controlling the speeds at which these activities occur. What are these controls and how do they act? The next few chapters, including this one, will provide some answers to this question.

The first hints that plants do have such controls over their activities came from observations familiar to us all. Who has not had house-plants bend towards the light directed at them through a window? Similarly, who is not aware of the fact that if a growing plant is laid on its side, in a few hours at most, its branches will begin to bend upwards and its roots downwards, in response to gravity?

The first recorded, scientific observations that plants could control their growth came in the 1870s. Theophil Ciesielski, a Polish scientist, discovered that when the very tip of a root (the root cap) was carefully removed, the rest of the root would no longer bend downwards in response to gravity. Ciesielski concluded, quite correctly, that the root cap was the place where a stimulus was produced which then caused the growing root below the cap to curve downwards after being placed horizontally. He suggested that this stimulus somehow caused the upper side of a root laid horizontally to grow faster than the lower side so that the root curved downwards as it grew.

We now know that Ciesielski had the right idea; the upper side does grow faster in a horizontal root than the lower side causing downward curvature. This is due both to a speeding up of growth on the upper side of the root and to a slowing down of growth on the lower side when a plant is laid on its side. What this stimulus was and how it had its effect was not understood in Ciesielski's day or for a long time after, however.

The first observations of the presence of a substance or substances in plants that affected growth were made by Charles Darwin with the help of his son, Francis, during his study of the curvature of plants towards light. We often forget that Darwin not only produced creative views on evolution and natural selection which helped to revolutionize biological thought but that he was also a gifted experimental scientist who made many contributions to science in addition to those for which he is rightly famous.

Working with seedlings of canary grass *(Phalaris canariense)* the Darwins found that if the very tip of a seedling shoot was covered with a tiny cap of blackened glass (to exclude light from the tip) the plant would no longer bend towards light directed at it from one side. If, on the other hand, the seedling was buried in fine black sand so that *only* the tip was exposed, curvature in the direction of light did occur. In fact, the *whole plant* bent toward the light even though only the very tip was left uncovered by the black sand. The Darwins also found that if even the last 3 millimeters (mm) of the tip of the seedling was cut off, no curvature toward light occurred. From these and many other observations, the Darwins reported in 1881 in a book, *The Power of Movement in Plants,* that:

> *when seedlings are freely exposed to a lateral light some influence is transmitted from the upper to the lower part, causing the latter to bend.*

In other words, it seemed as though the very tip of the canary grass seedling somehow sensed the direction of light and then passed on information to the rest of the seedling below causing it to grow in the direction of the light. Removal of just the tip of the seedling and the rest of the plant seemed to be rendered 'blind'.

Only many years after the Darwins was the true nature of this influence on growth finally worked out. By the 1920s, others had concluded that the stimulus produced at seedling tips was a chemical of some kind. But plants contain thousands of chemicals. Which one was responsible for transmitting the 'curvature stimulus' from the tip to the rest of the seedling? Was it only one substance or several acting together? Nobody knew.

One observation made by the Darwins many years before turned out to be centrally important in the search for the curvature stimulus. The Darwins had found that if just the very tip of a seedling was cut off, the rest of the plant failed to curve towards light. Fritz Went, a scientist working in the USA in the late 1920s, reasoned that it should be possible, therefore, to restore the ability to curve towards light to a seedling from which the shoot tip had been removed (decapitated) by adding back to it the particular chemical lost when the tip was cut off.

Went painstakingly separated from one another the chemicals he extracted from shoot tips and added them back to decapitated seedlings. After a great deal of patient work, Went found that there was, indeed, a substance that when added back to decapitated seedlings restored their ability to curve towards light. But the identity of the substance remained elusive until it could be isolated in larger amounts, not from plants, however, but from a totally unexpected source – human urine!

Fritz Kögl, working at the University of Utrecht in the Netherlands in the 1930s, set about trying to identify this elusive 'curvature compound' in plants. Kögl found that human urine, of all things, had in it a chemical which, when added to decapitated seedlings, restored their ability to curve towards light. From about 180 l of urine, provided by a local hospital, Kögl isolated 40 milligrams (mg) of crystals of a chemical which, when dissolved in water and added to decapitated seedlings, powerfully restored their ability to curve towards light.

The substance in urine proved to be identical to that found earlier by Went in seedling tips. Kögl called the compound *auxin* (the Greek word *auxein* means *to increase*). Soon, others showed that auxin was also responsible for the bending of roots towards and stems away from gravity when plants were laid on their side. Auxin, it was quickly realized, was a substance which fitted the definition of a hormone, a word used first by scientists who studied animals.

For a long time before Went's and Kögl's studies, it was known that animals produced compounds in very small quantities in one place and transported them to other locations in the body where they had their effect. For example, hormones are produced in the brains of humans and then move in the bloodstream to the sex organs where they influence reproduction. Many other organs of the body produce other, similar kinds of compounds in very small quantities which are then carried elsewhere before affecting some function of the body.

Auxin, too, is produced in one place in a plant (one example being a seedling tip) in very tiny amounts and then moves to other parts of the seedling (lower down the stem or even all the way to the roots) where it affects growth. Soon, it was realized that auxin was produced not just by a few types of seedlings but by *all* plants. It is truly a universal substance within the plant kingdom.

Plant scientists have adopted the term *growth substance* rather than hormone for compounds like auxin. They have done this because it is now realized that, although plant growth substances resemble animal hormones, they are not identical to them. It is thought best, therefore, not to use the same word to describe them.

The early investigations into the simple act of plant curvature towards light were the beginning of a journey of discovery of how *all* plants control their growth and development, a saga which has continued at an ever increasing pace to this day. Even at the time when auxin was thought to be the only plant growth substance (as we shall see, this is no longer true), it quickly became clear that its effect on the curvature of stems and, as was later realized, roots, was but one of many it had on a variety of different kinds of events in all plants.

For example, as long as the lead shoot of a plant remains intact its side branches grow more slowly. A clear example of this is the so-called 'Christmas tree effect'. The branches of the conifers usually used as Christmas trees are arranged in a distinct way; short near the apex of the tree and becoming ever longer from the top, down. Many plants have this pyramid-shaped growth form as long as the lead bud remains intact. As soon as that bud is cut off, however, the side branches begin to grow more rapidly. The branches take over if, for some reason, the apex is damaged or lost, through the attentions of browsing animals or from disease, for instance. Gardeners and horticulturalists remove lead buds from their shrubs, especially, so that the plants will grow more 'bushy' rather than tall and slender as is more likely to happen if the lead bud is left on a plant.

All indications are that it is the auxin produced in the lead bud and transported down the main stem which inhibits the outgrowth of branches. As soon as this source of auxin is removed the branches begin to grow faster.

At about the same time as the branch-*inhibiting* function of auxin was discovered, it was also shown to have root-*forming* activity. The commercial value of this was quickly recognized and put to use. Cuttings of a wide variety of plants will grow roots directly from newly cut surfaces if auxin is added to them. Leaves, pieces of stem or root, or even bulb scales, if treated with auxin, can be made to produce roots in places where roots grow only slowly, if at all, in a normal plant. Dipping the cut surfaces of slips (cuttings) into auxin solutions or powders has become standard horticultural practice. Today, there are dozens of preparations on the market designed for the home as well as the professional gardener.

Another property of auxins that has become economically important is their ability to trigger fruit formation *without* any pollination when added to certain plants. Tomatoes, for instance, are usually grown commercially in greenhouses where there are few insects and no wind to aid in cross-pollination. Tomato growers, therefore, resort to spraying auxin on their plants to induce fruiting and avoid the slow, frustrating task of pollinating each flower by hand.

Although apple and pear trees do not need auxin application in order to fruit abundantly, growers of these crops have used auxin for yet other reasons. One of the main sources of loss in apple and pear crops is premature drop of the fruit. From one-fourth to one-half of an entire crop can sometimes be lost if the fruit falls before it has matured and developed good color. Growers previously were faced with either harvesting before the best quality was attained or else risking a heavy premature fall. Spraying auxin onto apple or pear trees delays the fall of the fruit and increases the harvest.

Still other commercial applications of auxin take advantage of its activity in inhibiting, rather than promoting, growth in certain cases. For example, potato tubers can be treated with auxin and prevented from sprouting in storage; in this way tubers can be kept longer. Artificial auxins, such as the familiar 2,4-D (2,4-dichlorophenoxyacetic acid), have for many years been used as highly effective weedkillers of broad-leaved plants while having no effect on grasses. Under the right conditions, 2,4-D was, and still is in some cases, used on sugarcane, and maize crops as well as golf courses and lawns to keep down common weeds.

I have given this list of the effects of auxin to illustrate that, while the activities of plant growth substances parallel those of animal hormones up to a point, the parallel is not complete or exact. Hence, the need to give them a different name and look at them differently from their animal counterparts. Whereas an animal hormone is likely to control a single process or function (such as an aspect of growth or reproduction), a growth substance like auxin influences a plant in many different ways. As we have seen, sometimes it stimulates growth; under other conditions it inhibits growth. And it may carry out actions in any part of the plant not just at a particular location. Animal hormones often have precise targets (such as the sex organs for reproductive hormones) and have no effect anywhere else in the body. The plant equivalents have general and widespread effects on many aspects of growth and development. As we shall see, this is as true for the several other plant growth substances that we now know exist in plants as it is for auxin.

For two decades following its discovery in the late 1920s, auxin was the only natural plant growth substance widely thought to exist. Repeated attempts were made to explain how it was that auxin could control all plant growth and development. But as time went by it became increasingly obvious that auxin could not possibly be the only growth substance in plants; the conviction grew that there had to be more than just one.

One reason for this conviction had to do with tall and dwarf pea varieties. The former become very tall during a growing season while the latter always remain short. Auxin was known to increase growth so it was logical to assume that adding it to dwarf peas would make them grow tall. Not so, auxin had no effect on the growth of dwarf peas. So, the question became, why do tall peas grow tall? Or alternatively, what is missing from dwarf peas that tall peas have?

Ironically, the answer sought was already at hand at the time it was asked but nobody recognized it. In fact, at about the same time as auxin was being identified, discoveries were being made in Japan that would provide the answer to these questions.

A young scientist, Eiichi Kurosawa, had been investigating the reason for a very serious problem in *the* major crop in Japan. Rice plants infected with the fungus *Gibberella* grow very tall and spindly; much taller than normal. This excessive growth weakens the stems of the rice plants which then 'lodge' much more easily when battered by rain or high wind. The stems are too weak to stand up to heavy rain and wind and just fall over on top of one another. This, in turn, reduces the eventual yield, as it would in any crop. The Japanese have given this condition in rice the colorful name, 'foolish seedling disease'.

Kurosawa found how to make the stems of rice plants grow long and spindly without infecting them with the *Gibberella* fungus. All he had to do, he discovered, was to add some of the nutritious broth in which he had previously grown the fungus in his laboratory to the leaves of rice plants. He came to the obvious conclusion that the fungus produced some substance which leached out into the broth in which it had been growing. The active ingredient in the broth, however, proved difficult to isolate although its action was quite clear. Nonetheless, by the late 1930s, Japanese chemists had isolated an active, crystalline substance which they called *gibberellin A*. When it was later found that plants, not just fungi, *also* produced compounds very similar to gibberellin A it was soon realized that here was a second possible growth substance to be set alongside auxin.

Around 100 gibberellins have now been discovered; all have much the same chemical structure. One of the most pronounced effects gibberellins have is to change the rate of plant growth; one of their most dramatic effects is their ability to increase the growth rate of dwarf stems. Dwarf peas *can* grow tall; all they lack is enough of the right kind of gibberellin to stimulate their shoots to elongate more rapidly. Addition of gibberellin to a cabbage plant converts the 'head', which is really a dwarf stalk, into a stem 1.8–2.4 m tall. Plants like sugarbeet, which form a 'rosette' of leaves close to the ground, can be made to 'bolt' to great heights by a gibberellin treatment.

Incidentally, treating naturally tall plants with gibberellin does not affect their growth. They already contain enough of the growth substance and do not benefit from having more added to them.

But the dramatic effects of gibberellins do not begin and end with dwarf plant growth. Another event in the life of many plants is dormancy, a topic which is important enough to have a chapter of its own later (chapter 10).

In temperate climates in late summer, deciduous plants produce overwintering buds which do not grow out until the following spring. These dormant buds form even though the air temperature at the time they are produced by the plant is still high enough to allow the rest of the plant to continue growing for many more weeks. Seeds, too, often show dormancy. In some varieties of lettuce and tobacco as well as in many weeds, for example, seeds fail to germinate unless exposed to light. Many weeds will only begin to germinate if the ground they are lying in is disturbed (by cultivation or by burrowing animals, for example) and they find themselves exposed to light at the surface of the soil after being buried, perhaps, for years. Treatment of dormant seeds such as these *in the dark* with gibberellin will cause them to germinate. Treatment of dormant buds with gibberellin likewise can cause them to begin growing immediately rather than after an overwintering period.

The gibberellins are found in many different types of organisms. Each species of plant has at least a few of the gibberellins that have so far been detected. But they are also common among fungi, not just *Gibberella*, and in bacteria. All parts of plants contain gibberellins with the highest amounts found in seeds. Young tissues have more than old ones. In general, gibberellins are concentrated in the most vigorous parts of the plant, not surprisingly given their association with rapid growth.

The discovery of this second type of growth substance in plants was just the beginning of the search for growth-active compounds to add to auxin. Several more have now been discovered; no doubt more are yet to come.

As early as 1913 it was shown that when plants were wounded, with a knife for example, new repair tissue quickly formed at the cut surface, rather as a 'scab' forms when we are injured. Repair could be prevented by washing wounds with running water immediately after making a cut. The obvious conclusion was that the water was washing away something essential for healing. The unknown substance was called a 'wound hormone' for obvious reasons but its identity remained elusive.

Of course, the new tissue for wound healing is formed from new cells which replace those damaged during cutting. The idea arose, then, that perhaps a place to look for a so-called wound hormone would be where new cells were routinely produced in large numbers. The question became, then, where in a plant does rapid new cell production occur? One obvious answer was in the ovule during seed formation. Here, not only is the new embryo being formed but also all the other parts commonly found in a mature seed, all of which need new cells. In the 1960s a natural compound with a powerful influence on the formation of new cells was isolated and purified from maize seeds.

This class of growth substances became known as *cytokinins* (*cyto* = cell; *kinein* = movement or growth), a name which describes the ability of these substances to stimulate plants to rapid production of new cells. They were quickly found in all plants as well as in other kinds of living things like fungi and bacteria.

The most bizarre, unlikely plant growth substance is a compound that has been known about for centuries but was not suspected until relatively recently to play a direct role in plant growth.

The Chinese knew long ago that fruits would ripen more quickly in a room in which incense was burning. Puerto Rican pineapple and Philippine mango growers built bonfires to produce smoke that would synchronize the flowering of their crops. Illuminating gas (for example, coal gas, once used to light streets, homes and offices, especially in Europe) leaking out of pipes caused leaves to fall off shade trees in certain German cities (reported as early as 1864).

By the 1930s, there was a general but still vague understanding that all of these apparently unconnected effects were caused by one substance, the gas ethylene. That something which was a gas at normal temperatures could be a plant growth substance took much longer to gain acceptance. We now know that most, if not all, plants and plant parts produce ethylene naturally. Mostly, the amounts released are small but even at very low levels, the gas can affect plant growth and development.

The most dramatic effect of ethylene is on the speed with which certain fruits ripen. Apples, pears, tomatoes, and bananas, to give four familiar examples, produce much increased amounts of ethylene just as the fruits begin to ripen. This extra ethylene hastens the ripening process and can lead to one fruit influencing the ripening of another. The old saying, 'One rotten apple spoils the whole barrel', refers to the fact that if apples (or pears, tomatoes, and bananas) in a closed container have among them just one that is overripe, the excessive amount of ethylene produced by this one fruit, because ethylene is a gas, can spread throughout the container and speed up the ripening of the rest of the fruits. Spoilage of this kind can be minimized by refrigeration (but not in the case of bananas which are temperature-sensitive for other reasons), since low temperature slows down ethylene production, or by quickly removing the ethylene by increased ventilation.

Not all fruits produce excess ethylene. In grapes, cherries, and the citrus fruits, ethylene plays no part in ripening. Yet, as far as can be determined, all parts of all flowering plants, produce at least some ethylene, if not necessarily very much. Does ethylene, therefore, have several roles in plants other than the acceleration of ripening in some fruits?

Interestingly, numerous mechanical and stress effects also increase ethylene production in plants. These include: increased pressure on a leaf or stem; attack by fungi, bacteria, viruses, and insects; the waterlogging of the soil around plant roots; and the onset of drought soil conditions. And in all these cases, ethylene plays a direct role in the response to the mechanical damage or stress to which the plant is subjected.

For example, if a young seedling is pushing its way up to the surface of the soil, having just germinated from seed, it is not very strong. If the seedling finds its way blocked by hard, baked soil, for instance, it may then not be able to force its way to the soil surface. What it can do at this point, though, is to react to the mechanical pressure on it from the hard soil by producing more ethylene which causes the stem of the seedling to slow its growth upwards but expand its growth in thickness. In this way, the seedling becomes sturdier which, in turn, gives the bulk needed for a more determined push through to the soil surface.

The overproduction of ethylene by a plant at the point of attack by insects, bacteria, fungi, or viruses can lead to the death of the plant tissue immediately around where the invasion has occurred. The brown spots on leaves under attack from insects or disease organisms are the result of the planned death of the tissues at those points. Ethylene production helps speed up this response and provides some measure of defense against the further spread of the disease organism or the insect attack. Disease organisms are less likely to spread if the tissue surrounding them is already dead. They become isolated in a ring of dead tissue from which they cannot escape to invade the surrounding, still living, tissues of the plant. In this way, the plant attempts to build a barrier between the invading disease organism or insect and the rest of the plant. Of course, it does not always succeed in stopping the advance of the attacking forces but may slow them down enough to give other defense strategies time to work. The ethylene may also act by directly slowing down the growth of the invading disease organism itself (see chapter 14 for a discussion of plant defense strategies).

Ethylene, then, might be regarded as a growth substance which helps the plant to react to various stresses imposed from outside. Both in fruit ripening and in the last examples given, the end result is a speeding up of processes resulting in rapid maturation, and indeed sometimes the death, of the tissues affected. This view of ethylene is further confirmed by the fact that its production is often increased in flowers as they fade. After flowers have served their function they tend to wither away quite quickly in most plants. The rapid decline is hastened by increased ethylene production in petals.

Not that it is easy to provide such a simplistic view of this, or any other, plant growth substance. After all, ethylene also speeds up flower *formation* in some plants and promotes seed germination in others. In other words it is not just an inhibitor but has multiple effects on growth. In this, it also resembles all the other plant growth substances.

Attack by insects and disease organisms are not the only stresses to which plants are subjected. In the normal course of a day out in the hot sun, for example, plants may find themselves short of water or the victims of high temperatures from which they cannot escape by moving into the shade as mobile animals might do. Conversely, a plant might find itself exposed to sudden cold weather which has to be endured for longer (over a winter) or shorter (overnight) periods of time. It was, in fact, during the search for an explanation of the ability of perennial plants to form overwintering buds that another plant growth substance was discovered in the 1960s.

A group of investigators in Great Britain searched for an explanation of the fact that trees (their example was the sycamore) form their winter buds long before the end of the summer. The question was, how do the plants 'know' winter is approaching when the weather is still warm? Others in the USA and New Zealand were trying to answer the question as to what causes fruits to fall when they are ripe. These groups of investigators independently came to the same answer at about the same time.

Plants contain a growth substance which has come to be called abscisic acid (ABA) because of its influence on the process that leads to fruit fall (and, it so happens, leaf fall as well) called, *abscision.* This has turned out to be a rather unfortunate name since the effect of ABA is by no means confined to abscision alone.

In fact, the main purpose of ABA seems to be to slow down the growth of the whole plant, or any part of it, for various reasons. Thus, if a plant lacks water, it increases production of ABA. This, in turn, causes the stomata or pores in leaves through which water is lost by transpiration (see chapter 1) to close. In this way the plant can conserve whatever water is left inside its tissues. At certain times of year, more ABA is produced which aids in the production of winter buds. ABA produced in seeds can cause them to become dormant (see chapter 10) and remain so until the level of ABA in the seeds declines some time after they have been shed from the parent plant. Higher or lower than normal temperatures can lead to the production of increased amounts of ABA in all parts of the plant which causes, in turn, the growth of all parts of the plant to slow down. When normal temperatures return, the level of ABA in the plant decreases and growth again speeds up.

ABA, then, acts something like the brake on a car. When conditions surrounding the plant are not very favorable, or in preparation to survive adverse conditions in the future (as in the case of seed dormancy or winter bud formation), ABA acts to slow the pace of, or stop entirely, normal life processes until better conditions return.

The long and the short of it is, then, that plants have in them a number of growth substances which, either separately or working together, guide the progress of all aspects of their growth and development in response to variations in the surrounding environment. The five types of these compounds described here are not necessarily the only ones; others will surely be discovered as more detail is learned about plant growth. Recent investigations of substances in plants called jasmonates is one case in point. These compounds occur in many plant families and species, and play a role in promoting senescence (see chapter 17). Without growth substances, or 'hormones', plants would be 'blind' to the variations in their environment.

A plant's genetic program may be the ultimate control of growth and development but growth substances are a necessary additional level of day-to-day response to the unpredictabilities of climate and weather.

8

The time of their lives

Green plants growing in those areas of the world where there are definite seasons must synchronize their activities to suit natural variabilities in the weather. For example, it would not make much sense for a seed in temperate regions to begin germinating as soon as it was produced, say, late in the summer because rapid onset of winter would surely freeze and kill the fragile seedling before it could become established. Similarly, a tree or shrub faced with an approaching cold season must begin, in advance, to make provision to protect tender growing points and leaves within buds.

In another example, an annual herb needs to develop to a reasonable size before diverting its energies to forming flowers and then fruits and seeds. Leaves need to be first produced and then used to accumulate sufficient food reserves through photosynthesis before energy-consuming tasks like flower, fruit, and seed production begin. Such events must be timed to occur in sequence, synchronized with the seasons.

For these and many other reasons, plants need to have some way of sensing the passage of time; some way of measuring the lengths of days and the onset of seasons. Those who live in temperate regions know from their own observations that plants can do this. For instance, crocuses bloom in spring, roses during the summer and chrysanthemums in the fall. How do plants maintain this kind of predictability?

We might suspect that each type of plant has an internal genetic program which determines when it will flower. Each living organism has a genetic blueprint which distinguishes it from all others. Is this blueprint also used to determine exactly when each species of flowering plant will bloom? The answer surely is that this cannot be so since, for instance, by artificially setting the length of day and night in the greenhouses where chrysanthemums are growing they can be forced to flower at any time of year desired; commercial horticulturists do it all the time. We cannot say, then, that chrysanthemums will only bloom in the fall. That may be true if we grow them outside in natural conditions but it is not invariably the case. Grow them under artificial conditions where their environment can be controlled and changed at will and their development can be altered.

The fact is that plants have developed a variety of ways to meet the challenges of recognizing the passage of time. The rhythms of plant life are linked to cycles, such as day and night and the seasons, through a number of remarkably exact time-measuring mechanisms.

We have known for a long time that there are rhythms in plant growth. Androsthenes, historian to Alexander the Great, was one of the first to observe, nearly 2350 years ago, that the leaves of some plants take up different positions during the day than at night. For example, in some cases leaves come together at night, like hands held in prayer or greeting, then unfold again during the day. Other plants have leaves which bend closer to the ground during the night and are held up higher (more exposed to the light) during the day. These so-called 'sleep movements' are common not just among leaves but also in flowers. Carolus Linnaeus, one of the eighteenth century founders of modern botany who established the system of classifying plants by giving them binomial Latin names, worked out a floral clock based on the opening and closing of various flowers, which was reputedly accurate to within half an hour.

One of the most accurate individual plant timekeepers so far discovered seems to be the Malayan evergreen bush, shrubbery simpoh *(Wormia suffructicosa)*, which flowers every day of its life for half a century or more once it has reached full maturity. Its flower buds open at 3 AM; flower petals fall at 4 PM the same day; the fruits that follow are ripe in precisely 5 weeks; they split open to release seed at 3 AM on the 36th day after the flower buds first open. The South American rain tree *(Albizia saman)* folds up its leaves not only at night but also in cloudy weather, just as many sun-sensitive flowers do. For example, the snow gentian *(Gentiana nivalis)* closes its flowers whenever a cloud passes over the sun and reopens when sunlight returns.

All of these examples point to the fact that there are in plants daily as well as seasonal rhythms. In addition to flower formation and the sleep movements of leaves and petals, there are the regular variations in such things as growth rates of various organs (leaves, stems, roots, and flowers) which may grow faster or slower at different times of day; fluctuations in levels of growth substances which increase and decrease on a 24-hour cycle; the daily opening and closing of the stomata in leaves through which gases are exchanged with the atmosphere; the release of scents by plants which is correlated with nectar production; changes in the rate at which photosynthesis takes place (other than the obvious changes linked to the fact that photosynthesis occurs only during daylight); and variations in respiration, to name but a few of the many known.

But it was the intriguing problem of the sleep movements of leaves that, historically, led to the earliest clear knowledge of rhythms in plants and the fact that plants have in them something resembling a 'clock' which they use for keeping time.

More than two and a half centuries ago the French astronomer, Jean Jacques d'Ortour de Mairan, asked a crucial question about the sleep movements of plant leaves. He wondered whether the movements were caused by changes in the environment around a plant (daily regular day–night changes, for example) or whether a plant had some kind of internal time-measuring system.

Using the *Mimosa* plant, that he knew already to have very sensitive and predictable daily leaf movements, de Mairan made note of the change in position of leaves in light compared to deep shade. He discovered that the leaves of *Mimosa* did not have to be exposed to sunlight daily but continued to open and close even in continuous, near complete darkness. He concluded that there was a kind of clock somewhere *inside* the plant which ran independently of any changes in the surrounding environment and which guided the sleep movements of the leaves.

We know now that de Mairan was correct; plants do have internal, independent clocks. What he missed is that regular, predictable changes in the environment surrounding a plant *do* influence the way in which the internal clock operates. Over 200 more years would pass, however, before the idea that plants had an internal system for telling time would take hold. Even today, the links between these 'biological clocks' and the environment are by no means clearly understood.

We now have no doubt, however, that many, if not all, living things have internal clocks of some kind. In humans there are regulated rhythms in sleep (for instance, waking just before the alarm goes off, most of the time anyway!); in our ability to stay alert and make complex decisions; in varying hormone levels, heart rates, body temperature, excretion of urine, sensitivity to drugs, births and deaths (both of which occur *most* frequently between 2 and 7 AM and *least* often between 2 and 7 PM); and a host of other regular bodily functions. Biological clock rhythms occur in virtually all plants, animals, and fungi studied so far. Even some bacteria seem to have them.

Studies of these rhythms in humans have many implications for modern life. For instance, some human disorders like winter depression (a significant problem in those countries where daylight shortens drastically or disappears entirely for part of the year) can be cured or very much helped by light treatment given at the right time of day. A periodic exposure to light at no more than 3 percent of the brightness of sunlight can aid those with winter depression. This need for bright light seems to be connected somehow to the clock or clocks in certain individuals who can become seriously depressed without it. Air travel across time zones has a strong effect on the clock systems of passengers and it may take several days to adjust to a new time zone. Knowledge of biological clocks is exploited in some hospitals; the sensitivity of patients to drugs is known to vary in a regular 24-hour cycle. Thus, the lowest possible drug doses can be given at times of day when patients are most sensitive to them.

Incidentally, this same kind of knowledge about daily fluctuations in human physiology should be applied to shift workers, medical practitioners, pilots, and others but is more often ignored completely when setting work schedules. The question has been asked as to whether the Three-Mile Island and Chernobyl nuclear disasters in the USA and Ukraine, respectively, both of which occurred beween 2 and 4 AM, might have had different conclusions had more attention been paid to the biological clocks, and therefore the alertness, of workers on duty in the dead of night at such sensitive facilities.

The most important breakthrough in our early understanding of clocks in plants came in the 1920s. At that time, in Frankfurt, Germany, two scientists, Erwin Buenning and Kurt Stern, were investigating leaf movements in common bean plants which, like the *Mimosa* used by de Mairan about two centuries earlier, expose their surfaces to the sun by day and fold them vertically to a 'sleep' position at night. They were helped in setting up their experiment by another researcher, Rose Stoppel from Hamburg, who was responsible for looking after the plants each day. Stoppel found that when these movements were measured in a darkened room at constant temperature and humidity, the leaves were always at their maximum sleep position at the same time each day, between 3 and 4 AM.

Although this result pointed to the same conclusion that de Mairan had drawn, that is, that plants have an internal clock, Stoppel refused to believe that the plants were capable of keeping such accurate time by themselves. She certainly had no doubt that a clock existed in the bean plant. But, she became convinced that there had to be some outside influence which was somehow resetting this clock each day. She had no idea what this influence from outside the plant could be and called it simply *factor X* reasoning that it had to be something other than light, temperature, or humidity, all of which she thought she had under strict control in her experiments. In this belief she was mistaken.

Stoppel was a person with a methodical way of working. One of her regular habits was to water her bean plants at nearly the same time each day. In order to see to do this in the darkened room in which the plants were housed she would turn on a flashlight covered with dark red paper. The common belief at that time was that red light had no effect on plants. Stoppel, therefore, thought that she could use a dim red light in the darkened room without affecting the bean plants. This turned out to be the kind of serendipitous error that all scientists wish they might make once in their lifetime; an error which leads to an unexpected, very important discovery.

Stoppel eventually returned to Hamburg without discovering the identity of factor X. Soon after, Buenning and Stern moved the bean experiment from where it had been located to Stern's potato cellar in an attempt to obtain better temperature control. Like Stoppel, they found that, indeed, the maximum sleep position of the bean leaves always occurred at the same time every day. What surprised them, though, was that the maximum was not between 3 and 4 AM as Stoppel had found but about 8 hours later, between 10 and 12 AM. Why the discrepancy?

Buenning and Stern quickly recognized that the key to the dilemma was the dim red light they, and Stoppel, had used to find their way around the darkened room in their daily routine of watering the plants. The potato cellar used for growing the beans was far away from the laboratory where Buenning and Stern worked during the day so they waited until late afternoon before carrying out this daily task. Stoppel, on the other hand, had done her watering in the morning, about 8 hours earlier than Buenning and Stern.

The daily, brief exposure to the dim red light from the flashlight was enough to reset the bean plants' clock daily so that 16 hours later leaves folded into their maximum sleep position. Since Stoppel watered in the morning, 16 hours later was around 3 or 4 AM; for Buenning and Stern, who watered in the afternoon, it was between 10 and 12 AM.

In this example, the brief exposure to dim red light was probably interpreted by the bean plant as the signal it would normally receive at the dawn of a day. The transition from dark to light at dawn each day resets the biological clock controlling sleep movements. Of course, in regions of the world away from the equator, dawn comes at a slightly different time each day, year round. Resetting the clock in response to a reference point such as dawn allows the plant, then, to measure internally the time to the middle of the next night, 16 hours ahead. Of course, Stoppel and, later, Buenning and Stern did not vary their simulated 'dawn'. Therefore, the next point of maximum folding of the leaves in sleep was always at the same relative times in the experiment before and after the move to the potato cellar.

Quite unexpectedly, red light turned out to be Stoppel's factor X! Plants are indeed affected by red light, a discovery which led to one of the most imaginative and intuitive biological investigations ever undertaken, as we shall see in the next chapter.

Buenning and Stern went on to make another unexpected, important discovery. When the use of red light was eliminated and the bean plants grown in truly complete darkness all the time the period of the leaf sleep rhythm of opening and folding up no longer took place on a strict 24-hour cycle. In fact, the 'daily' cycle was about 25.5 hours, and was soon, therefore, completely out of phase with the cycle of light and dark outside the root cellar. The rhythm was no longer synchronized to a 24-hour day but had become, instead, 'free-running', completely independent of the light and dark cycles going on in the environment outside the potato cellar. A 24-hour cycle could be reimposed on the free-running rhythm by once again briefly exposing the plants to dim red light at the same time each day. Whenever that was done, the maximum sleep position of leaves was reached, once again, 16 hours later. Because these free-running rhythms were close to being 24 hours in duration, but not exactly so, they became known as 'circadian' (each complete turn of the cycle lasting *circa* = about; *diem* = a day).

We now understand that in constant conditions, such as continuous darkness (or constant light for that matter), all rhythms linked to regular day and night cycles, not just the sleep movements of bean leaves, drift away from an exact 24-hour cycle. They move to some value which is usually somewhat shorter (maybe 21 or 22 hours) or a little longer (25 or 26 hours) than a regular day. These free-running, natural circadian rhythms can be brought back onto a 24-hour cycle by exposure to some regular, repeating signal such as a dark to light transition at dawn or, in some cases, dusk. Furthermore, the internal clock which measures this passage of time is virtually independent of temperature change, as might be expected. A clock of any kind would be quite useless if the rate at which it operated changed as the temperature increased and decreased from day to night, for example. One of the essential characteristics of a clock is that it keeps constant time no matter what is going on around it. This is no less true for the biological variety than for any other type of clock.

Thus, circadian rhythms linked to 24-hour light–dark cycles show a number of interesting features. The rhythms originate from patterns in the environment that repeat over and over again. If these repeating patterns of light and dark are removed and constant conditions imposed, the rhythms drift in relation to solar time, either gaining or losing time depending on whether they drift to shorter or longer cycles than 24 hours. Under natural conditions, the internal clock can be synchronized in most living things to a true 24-hour period by the repeat signals derived from regular changes of light and darkness in a normal day.

One of the more important of these signals seems to be the dark to light transition at dawn. Such signals have been given the name *Zeitgeber* (the German word for 'time-giver'). Unless reinforced by regular Zeitgebers, most circadian rhythms will drift to some cycle length other than the environmental one. After a further while if a light–dark stimulus is not given even the free-running rhythm will become progressively weaker and eventually peter out entirely. In other words, some stimulation from the environment is crucial to the maintenance of these rhythms even in their free-running form.

Light and dark are not the only Zeitgebers, of course. In ocean tidal zones, marine animals and plants synchronize their feeding and reproductive activities to the ebb and flow of the tide. If removed to bodies of water without tides, some marine forms will continue this imposed pattern of behavior, just as in the case of any other rhythm, as though the tide was still there. Eventually, as in the case of other rhythms, the activity will peter out unless the organisms are stimulated by further exposure to a tide. From time to time, claims have been made that there are other rhythms associated with the lunar month but evidence is not yet very convincing or free from alternative explanation.

Resetting the clock using something like the transition from dark to light at dawn (or dusk would do just as well) makes sense since many rhythms are linked to daytime or nighttime activities. Because away from the equator, the lengths of day and night vary with the seasons, some signal linked to the fact that the time of sunrise varies each day is essential.

Thus, resetting the clock in synchrony with the seasonal changes in daily light and dark periods makes sense. In the case of bean leaves, for instance, it is important that the leaves be in their maximum open position around the middle of the day when they can carry out the most efficient photosynthesis. It would be no use to have the leaves in their maximum sleep position during daylight hours, for example. In order to keep track of the start of the day, however, some mechanism is needed to inform the plant when the transition from dark to light is occurring each day throughout the growing season so that the reaction of the leaves, in the case of bean, can be synchronized with daylight hours. The light–dark transition Zeitgeber is widely used by plants to set internal clocks not just for leaf movement but for many other activities as well.

And this also provides an answer to the question as to why only a few environmental fluctuations are favored as Zeitgebers. Why light–dark and tidal changes but not temperature or humidity, for example? It is true that there are many repetitive changes in most environments other than light and dark, and the tides. In fact, change in its environment is the only thing that an organism can count on. Almost nothing is really constant. Wind velocities, temperatures, light levels, and humidities all can change, either rapidly or slowly and sometimes very drastically. All fluctuate daily. In reality, though, only three of the cycles mentioned above are regular enough to allow organisms to adjust to them by developing inbuilt mechanisms to deal with them. Changes in temperature, wind velocity, light levels during the day (for instance, due to changing cloud cover), and humidity are all too unpredictable to form the basis for regular reactions by organisms.

Certainly, in temperate climates, temperatures, to take one example, vary regularly from season to season. We can be fairly certain that it will be warm in summer and cold in winter. But those of us who live in places where there is any kind of seasonal change in climate know how unpredictable temperature variations can be, daily and from one time of year to another. Yet accuracy is one of the fundamental requirements for any clock system. Thus, daily changes in light and dark, annual fluctuations in daylength, and the tides, are the only environmental variations which are sufficiently reliable to be used to set and reset clocks.

We have spoken here mainly of circadian rhythms but there are other rhythms besides that are not even close to being one day in length. Although they are not visible to the naked eye, since they occur rather slowly, some movements of parts of plants have regular cycles which are much shorter than one day, perhaps only a few minutes or not much more than an hour. For example, as plant shoots grow upwards, they do not move ahead perfectly straight and steady; they sway in gentle circles as they lengthen. In peas, one complete cycle of this circular swaying motion takes about 77 minutes. As leaves expand to their full size, they sway and rotate in repeatable patterns which have rhythms taking just a few minutes.

In other cases, there are some rhythms which are much longer than 24 hours. Four-day, rather than 1-day, rhythms have been found to persist in the growth and reproduction of some fungi. The reason for rhythms with periodicities other than 1 day may be lost in the dim and distant past of these very ancient living things. For example, it is believed that two and a half billion years ago, the lunar month might have been 40 to 45 days long with each day being only 5 hours in length. Others have suggested a 425-day year for our planet 600 million years ago. Perhaps rhythms differing widely from a 24-hour period are remnants of earlier times when daylengths and other regularly cycling conditions were not as they now are. On the other hand, maybe these particular, strange living things just have never quite got it right or are on a different cycle for reasons we do not understand. We may never know the answer.

A few cycles of one or several *years* are also known. The sensitivity of bean plants to red light has already been dealt with in some detail but there is other evidence that this sensitivity is not the same at all times of year. Thus, the bean is more sensitive to red light during its normal growing season than at other times of year. The flowering of some bamboo species have cycles of 30 or 40 years. One particular bamboo species native to the mountains of Jamaica carries the process of growing to a remarkable extreme: exactly 32 years after the plant germinates from seed, it flowers once and then dies. And this pattern seems to be independent of the environment because transplanted to any other part of the world, the plant still blossoms on schedule at the age of 32 years, no earlier and no later.

To make matters even more confusing, we also now know that a single organism can have more than one free-running cycle which suggests, in turn, more than one clock per individual. For instance, such things as periods of activity and calcium excretion in humans both have natural cycles 33 hours in length whereas body temperature, urine volume, and release of potassium into the urine, all have free-running cycles of about 25 hours (of course, all can be made to conform to a 24-hour cycle by the appropriate light–dark treatment of a normal day). So, humans seem to have more than just one clock. Other organisms seem to have as well.

The really difficult questions about clocks is what and where they are in the plant and how they work? It may be obvious that plants have clocks but by what mechanism do they actually control what the plant does. Is there more than one kind of clock in any one organism? Can they be different in different organisms?

So far, there is no evidence for different kinds of clocks in different organisms. We all seem to have clocks based on the same mechanism. Location within the organism is a mystery still. There is general agreement that each and every living cell of an organism has its own clock. Much evidence suggests that each cell of a plant or animal reacts as though it had a built-in clock mechanism. But to say that this is so just begs the question of where within the cell, in turn, the clock is located? No clear answer to this quandary yet exists.

Some progress is being made, however, in the matter of the location of the clock in the cell through the fact that mutants have been found in which the clocks are changed. In some species, mutants have appeared which have certain rhythms that are different from the normal in ways that are inherited and passed on from one generation to the next. These mutant forms could be important in attempts to find out where a clock is located. If it can be discovered what has been changed in these mutants, perhaps the clock location can then be pinpointed.

There is no question that organisms have ways of measuring time. Many of these clocks are circadian and linked to daylength since solar cycles form such a reliable basis for timekeeping. Others may be longer or shorter than a day in length, sometimes by quite a bit. In all, light often plays a role in setting the rhythms. Thus, a major question becomes how organisms sense light and dark? Obviously not the same way in all cases. In animals like ourselves, light can be seen with the eyes. But what about plants? How do they 'see' the light which resets their clocks? Remember, the amount and quality of light needed for resetting is very slight. Dim red light from a flashlight will do. Thus, light recognition by plants cannot be by the same mechanism as in the case of animals like ourselves; yet there has to be a way. The answer to that question involves one of the most imaginative and intuitive biological investigations ever undertaken. We shall deal with it in the next chapter.

Finally, despite the concentration here on circadian rhythms and the establishment in plants of the existence of daily clock mechanisms, it is arguable that seasonal rhythms are, in fact, by far the most important to plants overall. At the beginning of this chapter, I gave examples of the regular seasonal flowering of different types of common plants. Other long-term activities are such things as the production of buds, the development and fall of leaves, especially in deciduous trees, and the preparation for winter. Gardeners know well that these activities are not markedly affected by weather. A very cold snap may slow down leaf or flower bud opening, for example, for a time but if bad weather persists buds will open eventually even if the flowers or leaves they contain then freeze. Conversely, an unseasonal warm period can bring plants prematurely into leaf or flower.

Seasonal variations in the times of opening of leaf and flower buds are, however, usually small, often only a few days either way from year to year. Therefore, plants must have ways of setting rhythms which also accurately measure the onset of seasons. How this is achieved is the subject of the next two chapters.

9

A dash of seasoning

In the previous chapter, we saw that plants have ways to use the daily transitions from light to dark at dawn and dusk as Zeitgebers, signals to reset their internal biological clocks regularly. This information about whether it is in darkness or in the light can be used by a plant to time when to carry out repeated daily functions, like changing the position of leaves in daylight versus the night; produce daily bursts of perfume to attract pollinators; and open and close flowers for pollination at certain times of day.

Plants can not only use day to night transition to time repeated functions from day to day but also to measure the seasons of the year reliably. After all, down the centuries human communities have used the production of leaves, flowers, or maturing fruit by local plants as signals for seasonal activities such as when to begin planting crops. The agriculturalists Wightman W. Garner and Henry A. Allard (more about whom shortly) expressed it this way:

> *One of the characteristic features of plant growth outside the tropics is the marked tendency shown by various species to flower and fruit only at certain times of the year. This behavior is so constant that certain plants come to be closely identified with each of the seasons, in the same way as the coming and going of migratory birds in spring and fall.*

As I said at the beginning of the last chapter, in the northern hemisphere we have neither crocuses in September nor chrysanthemums in May, not under natural conditions, anyway. Plants do seem to have a way of measuring time so that their most important activities, like seed germination, flowering, and fruit set, occur at particular times of the year.

Having said that plants do carry out particular functions at certain times of year, it is also true that we can manipulate and change when in the life of a plant these major events will occur. Thus, we can arrange to have chrysanthemums produce flowers at just about any time of year we choose now that we understand how to do it. So, it may be true that plants can measure the seasons and do so when they are growing in the open either under cultivation or in the wild. But it is also true that the measuring mechanism inside the plant can be artificially manipulated.

The question to be answered here is how plants 'see' light and dark and how this information about their environment is used to measure long periods of time like the seasons of the year?

In temperate climates, there is a definite pattern of plant growth arranged around a yearly period of inactivity, winter. The pattern is seen most clearly in annuals, which live within a regular yearly pattern, growing and reproducing during favorable weather and spending a resting period, winter, as seeds.

But other kinds of plants also have distinct patterns of growth. For example, herbaceous perennials produce annual stem growth and then stop well before the approach of winter; deciduous shrubs and trees lose their leaves in the fall and begin preparing for that event well in advance of the cold of winter. In all cases of periodic growth, some signal causes the plant to go from full-scale, active growth to near complete shutdown in a matter of days or weeks.

In the tropics, where variation in light and dark is absent or only small, plant responses are different from temperate climates. Tropical trees, for instance, are often laws unto themselves. If they rest at all, which means losing leaves for a short time, they are apt to do so individually, not synchronized with their neighbors even when of the same species. One species may flower in a cycle of a little less than one year; another in a few months greater than a regular annual cycle; some do so continuously, often one branch at a time; and yet others bloom only once every few years. This *laissez-aller* behavior is consistent with the congenial, nearly constant environment typical of the tropics in which timekeeping is of little importance. Plants that normally grow in stricter temperate climates can behave strangely when transplanted to the benign tropics where there are no seasons. Pear trees planted in Indonesia, for example, became evergreen, each bud having its own cycle of activity not synchronized with others on the same plant.

There is very little evidence that plants are aware of the passing of a period of time as long as what we call a year. In any case, we would find it difficult to show that a plant was, somehow, measuring such a long span of time. Yet, undeniably, many plants do behave as though strongly linked to the seasons. Somehow, they are capable of marking these changes. What is it that they measure? The beginning of an answer to that question came through the study of flowering in plants.

In 1910, a doctoral student in Paris, Julien Tournois, began studying flowering in Japanese hops. As a typical student he was impatient to hurry his study ahead so he planted seed in February, 1911, instead of at the usual time later in the spring. To his surprise, the plants flowered in April, even though they were only 15 cm tall. The hop is usually a long, straggly, twining plant which produces a meter-long stem before blooming in July.

When in the following year, plants that were started from seed in January flowered in March, Tournois began to think that somehow daylength might be affecting the timing of flowering. He wondered whether deliberately shortening the length of a day during a normal growing season would produce the same result. So, in April, 1912, and the year after that, he sowed hops in his garden under glass. Those plants grown in natural cycles of day and night grew tall and spindly, and produced nine pairs of leaves before flowering in early July, as expected. Other individuals covered with dark screens for all but 6 hours a day grew just taller than ankle height, developed only three or four pairs of leaves but flowered around June 20, earlier than the plants growing in natural conditions. Tournois came to the conclusion from all this that:

> *In the Japanese hop, a decrease in daylength during the normal growth period provokes some floral reproduction.*

After seeing the same result with hemp, Tournois' conviction that his plants flowered when still young because they were exposed to short days from germination onwards became even stronger. What is more, he came to the additional most important conclusion that it was the length of the *night,* not so much the shortness of the day, that was the deciding factor in when his plants flowered.

With these deductions, Tournois was on his way to making a highly significant contribution to our understanding of the control of seasonal rhythms in plants, a promise which, sadly, would never be fully realized. He was killed in action in World War I.

This, and other early work on the control of flowering in plants was, at first, quite unknown to the investigators, Garner and Allard, who began their work at the United States Department of Agriculture in the period just prior to 1920. They wondered why the variety of tobacco called Maryland Mammoth was delayed in flowering when it was grown near Washington, DC. There, it flowered so late in the season that its seed did not mature before the onset of winter. It was important to discover the reason for the delay in flowering because Maryland Mammoth had an enormous commercial potential at the time. This variety had been noticed first as a mutant in a field of normal-sized tobacco around 1906. The plants were conspicuous because they grew to a height of about 2.5 m and produced up to 100 leaves for harvest. Commercially, this was a big advance over other tobacco varieties then available.

Garner and Allard tried growing their plants under many different conditions and explored countless, fruitless ideas before discovering why it was that flowering in their tobacco began so late in the season.

The curious fact was that at the most critical time of year for triggering flowering in Maryland Mammoth, the days were too long and the nights too short at the latitude of Washington, DC. Garner and Allard discovered that this variety would not even begin to form flower buds until the length of daylight had shortened to between 10 and 12 hours. In daylight hours longer than this, the plants went on forming new leaves indefinitely, never flowering. In the Washington area, the days do not become as short as 10 to 12 hours until late in the summer. By then it is too late for Maryland Mammoth to produce mature flowers and then set seed before winter descends.

Garner and Allard went on to test this discovery of a link between flowering and daylength with many other species of plants. Eventually they found that, in general, plants could be placed in three major groupings: those that needed days of less than about 14 hours in order to set flowers (called 'short-day' plants); those that needed periods of daylight longer than about 14 hours, but not continuous light (called 'long-day' plants); and those which were not choosy about daylength (called 'day-neutral' plants). Garner introduced the word *photoperiodism* to describe this response of plants to differing lengths (periods) of light (photo-).

Many more examples have now been added to the groupings suggested by Garner and Allard. Examples of short-day plants would be annuals originating in the tropics, where short days are normal all year round, and most species found in temperate areas of the globe which flower either in spring or in the fall when daylength is shorter than in mid-summer, the late-blooming chrysanthemum being one of the best examples. The long-day plant group includes examples like wheat, potato, lettuce, sugar-beet, and other species of temperate regions that flower in the long days of mid-summer. Among the day-neutrals are plants like maize, tomato, most weeds (dandelions, which seem to bloom at every available opportunity, being a superb example), and the many tropical plants which produce flowers and also leaves the year round, having in fact no specific annual rhythms.

Thus, first from Tournois', and Garner and Allard's astute observations, and then from many other studies besides, it became clear why certain plants flourish in some latitudes and are just about absent from others. In the tropics, for example, where days remain close to 12 hours long, it is only short-day and day-neutral plants that succeed in flowering. No long-day plants could possibly flower where daylengths never exceed about 13 hours. In very high latitudes, such as the Arctic, we find only long-day and, again, day-neutral plants. Short-day plants do not flourish there because, although short days eventually come to such regions, they do so only briefly and too late in the season to allow flowering and seed formation before frigid weather and total darkness descend.

A particularly representative and widespread example of a short-day plant is ragweed *(Ambrosia artemisiifolia)* in which flowers begin to develop when the daylength has shortened to around 14.5 hours. At the latitude of Washington, DC, this daylength occurs around July 1, so the plant has plenty of time to flower and shed pollen by mid-August. Then, the seed sets and is released in good time before winter. But ragweed is much less common in more northerly latitudes, such as in Canada, for instance. There, the long summer days do not shorten to around 14 hours until early August. Ragweed forming flowers after that time would nearly always be killed by frost before seed matured.

Conversely, of course, a plant that thrives in more northern climates may fail to flower further south where spring and summer days are shorter. Daylengths may never be long enough for some species. For example, some rock-garden plants, like certain *Sedum* varieties, require days of 16 hours or more for flowering and will bloom in northern latitudes but not further south. Spinach would never flower or reproduce by seed in the tropics because it has to have a minumum of 14 hours of daylight for a period of at least 2 weeks. Such a combination of daylength and duration never occurs in the tropics.

Garner and Allard's explanation of flowering based on photoperiodism also provided a reasonable explanation of why all individuals of a certain variety will flower at the same time in a particular location even though they may be planted at different times during the growing season. Exposure to the correct daylength for a few days or even weeks is usually necessary to trigger flowering no matter when a certain species is planted. If short-day plants are kept indefinitely in, say, daily cycles of 16 hours of light and 8 hours of dark (long days) they will never flower. Similarly, long-day plants kept indefinitely in cycles of 8 hours of light and 16 hours of dark (short days) also will continue with leafy growth and never bear flowers. But then, after a few weeks of growth under the wrong regime of light and dark for flowering, if the two groups of plants are quickly switched so that they are now in correct day and night lengths, production of flower buds goes ahead in all individuals, at the same time, no matter whether they were originally all planted at the same time or not. Plants only need to be exposed to the right light and dark regimes for a minimum period of time (a few days or maybe weeks in some cases) after which they will go ahead and flower.

The discovery that the flowering of plants could be controlled by artificially setting daylength quickly led to major advances in the marketing of some major horticultural crops. One of the first to be exploited in this way was the chrysanthemum which in the 1920s was simply a fall-blooming garden plant in northern latitudes. But Allard was able to advance flowering of this species from mid-October to mid-July by darkening the greenhouses where the plants were growing. He began growing his chrysanthemums indoors in the spring. As the days grew longer in May and June, he darkened the rooms housing his plants to keep daylengths short, thus advancing the date of flowering by about 3 months.

In the 1930s, horticulturists began following Allard's lead and found they could produce chrysanthemum blooms at will at any time of year by manipulating daylength. To keep cuttings in just leafy growth until flowering was desired, they took to extending the short days of winter as well by using artificial light in their greenhouses so that chrysanthemums could be produced for the Christmas market. Such artificial manipulation of daylength is now a routine procedure for many horticultural crops.

Garner and Allard presented such persuasive evidence for photoperiodism that enthusiasm for it spread around the world. Sometimes their ideas were useful in the most unexpected ways. In one early example, in 1947, their discovery rescued farmers in Trinidad from a disaster not of their own making. Early that year, an oil company had erected a 7.6-m flare-pipe (to burn off unwanted gases) at the edge of a swamp padi (rice) field. To the dismay of farmers, the crop around the flare failed to flower as usual in October, even 100 m away where the light was very dim. Investigators called in to try and solve the problem noticed that some plants did flower in an area shaded from the flare by a large tree. From this they reasoned that light from the flare had extended the daylength so much that, being a short-day plant, individuals exposed to the extra light would not flower. When the oil company put out the flare, the crop again flowered normally.

Of course, once the idea had taken hold that daylength could control flowering in plants, there arose a curiosity about how a plant could possibly see light. In addition, there was the growing amount of information about how light and dark affected the biological clocks in plants (see chapter 8). The same question about how plants sense light arose there, too. It is all very well to say that plants use light and dark to their advantage but how do plants convert the fact of day and night, light and dark, into actions like when to flower? Attempts aimed at answering this question have provided brilliant insights into plant growth and development until today an entirely new branch of science has blossomed. We realize now that many aspects of plant growth are under the influence of light, not just flowering and circadian rhythms.

A critical breakthrough in understanding how plants see light was made in the late 1930s. Using cocklebur *(Xanthium strumarium)*, a short-day plant, it was discovered that the triggering of flowering was linked, not to the length of the light period each day, but to the length of the period of darkness. It became clear that short-day plants should really be called long-night plants instead. If a plant such as cocklebur was kept in short days (say, 8 hours of light and 16 hours of dark in each 24 hours) it flowered successfully. If, however, the 16-hour night was interrupted midway through by a short exposure to light, the plants would not flower. The exposure to light during the night did not have to be very long, a few minutes was enough; the light did not have to be very bright, something a little brighter than moonlight would do.

Here, then, was a puzzle. Very short periods of light in the middle of the night seemed to signal the plant that it was not in short days and long nights anymore but long days and short nights. A long, uninterrupted span of darkness seemed essential to sending the signal to the short-day plant, cocklebur, to flower. Any interference with this critical period of darkness seemed to block the long-night/short-day signal needed by short-day plants for flowering.

The opposite of a light interruption of the long night favored by short-day plants had no effect on long-day plants. If a plant which required long days and short nights in order to begin flowering had a short *dark* period imposed on it in the middle of its long day, flowering still went ahead. Interrupting the period of light with a short period of darkness was ineffective; the plant still saw itself as being in long days.

However, keeping a long-day plant on short days and then giving a few minutes of light midway through the long night *did* cause flowering. Once again, interruption of the night, not the day, was the critical point. Shortening a long night triggers flowering in long-day plants but prevents flowering in short-day plants. Just as short-day plants might be renamed long-night plants, so long-day plants might be called short-night plants instead.

So, the conclusion drawn by Tournois many years earlier in his experiments on Japanese hop was fully confirmed. Night length, not daylength, is the decisive factor in determining whether or not long- and short-day plants will flower. The importance of this early observation was made even greater when it was realized that knowledge of the effects of small amounts of light on plants in the middle of their night periods was the key to answering the question of how plants see light.

Exactly what goes on during the night in short- and long-day plants that is so critical to flowering is not yet completely understood. Even so, much has been learned about the control of plant growth in general, not just flowering in particular, by attempting to understand the influence of giving short doses of light to plants in the middle of the night.

The next critical advance was at first a puzzling one. One obvious next step in investigating the effect on flowering of short periods of light during the night was to ask whether there was any particular kind of light which was more effective than others in promoting or preventing flowering? Was blue light better than green or yellow, for example?

The answer again came from the use of the cocklebur plant where it was discovered that flowering could be prevented if short periods of *red* light were given midway through a long night. Long-day plants, similarly, could be made to flower if short bursts of red light were given midway through their long nights. These results were at first mystifying. Why red light? The answer came first, not from further investigation of flowering, but in other ways.

Back in the nineteenth century the discovery had been made that exposure to light improved the germination of certain kinds of seeds. The stimulation seemed to occur most often in the case of plants which produced small seeds. Small seeds have only very limited amounts of food reserves stored in them. Therefore, if they germinate when buried deep in the soil, there may not be sufficient food reserves to nurture the young seedling until it breaks through the surface of the soil and begins producing its own food by photosynthesis. Small seeds tend not to germinate until some activity around them causes them to be carried up onto or close to the soil surface where light can reach them. An animal (such as a rodent or an earthworm) burrowing into the soil and turning it over in the process can cause seeds to be thrown up onto the soil surface. Cultivation of soil by gardeners or farmers can cause the same thing to happen. As soon as such seeds sense that they are now near the surface, in the light, and not buried deep in the soil, in darkness, they begin to germinate.

Many seeds are noted for this kind of response which is why newly cultivated soil seems to become 'weedy' so quickly. Weed seeds may have been lying dormant deep in the soil, possibly for years. 'One year's seeds, seven years' weeds' is an old gardeners' saying, and weeds are past masters at spreading their germination over many years as and when their seeds are turned up to the soil surface during cultivation. The fantastic display of annual poppies 'In Flanders Field' during World War I, when shells churned up the soil, was another dramatic example of the ability of weeds to germinate when suddenly exposed to light. It has been estimated that there are up to a billion weed seeds per hectare of cultivated land waiting in the moist soil for the light that will release them from their dark-induced slumber.

One type of cultivated seed which requires light for germination is lettuce (not all varieties, but several, nonetheless). Lettuce has a small seed which remains dormant in the soil unless given adequate exposure to light after it has been planted and has become swollen with moisture (dry seeds will not respond to light). In 1934, Lewis H. Flint was hired by the United States Department of Agriculture with the task of finding out what kind of light would most effectively break the dormancy of light-requiring lettuce seed. Flint soon discovered that a mere 4 seconds of exposure to full sunlight prompted swollen Arlington Fancy lettuce seed to sprout.

About this time, Flint was joined in his work by the physicist, Edward D. McAlister. Together they set about answering the question as to what kind of light caused lettuce seeds to germinate, the same question as Garner and Allard had asked about the effect of light on flowering. The answer turned out to be the same in both cases.

Flint and McAlister found that germination of lettuce seed was increased to the greatest extent by red light, just as had been found with flowering. Not only that but they made another crucial discovery. They found that not only did red light increase lettuce seed germination but that infrared light (of the type you might see in bathroom heat lamps or in the dull glow from electric fires) *prevented* it. They found they could switch germination on and off by repeated treatment with red and infrared light. If they treated moistened seeds with red light, germination occurred; if the red light treatment was followed immediately by infrared, the effect of the previous red light treatment was canceled out. Eventually, it became obvious that the seeds seemed to have a memory and reacted to whichever type of light treatment was given last. In other words, Flint and McAlister found they could give repeated alternating red and infrared light treatments; if the last type of light seen by the seeds was red, then germination was promoted and if infrared, germination was inhibited.

The question then became whether the same was true of flowering? And the answer was a very clear 'yes'. If short bursts of red light were given to the short-day plant, cocklebur, in the middle of a long night, and followed at once by infrared light, the plants ignored the fact that they had been exposed to red light and flowered. It was as though the plants had seen no light at all. Red light given to long-day plants in the middle of a long night, as we have seen earlier, would normally be a signal to the plant that it was in a short, not a long night. This signal was also reversed by infrared light given right after red light treatment.

These unexplained, puzzling observations on seed germination and flowering stimulated the curiosity of two other investigators, Harry Borthwick and Sterling Hendricks, who then began studies which were to lead to a discovery that continues to inform and enrich our understanding of plant growth and development right up to the present day.

Between about 1945 and 1960 in the USA, Borthwick and Hendricks looked at the effects of red and infrared light on several aspects of plant growth and development, not just one as previous investigators had done. They repeated and greatly improved the conclusions of Flint and McAlister on the effects of red and infrared light on seed germination. They did the same with the work of Garner and Allard, and others on the influence of red and infrared light on flowering. And they, themselves, investigated a third effect of red light on plants.

In many kinds of plants (peas are a good, common example), when seeds are germinated in the dark, the young seedlings produced have long, spindly stems and tiny leaves which are yellow, not green. The word *etiolation* is used to describe this condition. We see this ourselves in plants grown with insufficient light in our homes or our gardens. Most plants grow in a 'leggy' fashion if not given enough light. Seeds germinated in complete darkness show this condition in the extreme. Shoots produced in the dark are long, thin, and weak; leaves are tiny and pale yellow.

Exposure of such sickly, weak-looking seedlings to a few minutes of sunlight causes an amazing transformation to take place even if the seedlings are returned to darkness immediately. The most noticeable change is in the leaves which begin to turn green and to grow to normal size within a few minutes of exposure to sunlight. More slowly, the long spindly stem begins to slow down its growth in length, and become thicker and sturdier. Many other, less obvious, changes in the growth form of the plant also occur at the same time.

The rapid, spindly growth of seedlings which typifies etiolation is an adaptation many plants have developed. After all, most seeds usually germinate in the dark, deeply buried in the soil. By no means all types of seeds germinate only when exposed to light although, as mentioned earlier in the chapter, many do. Many others germinate in complete darkness well below the soil surface. The smartest things for a seedling in this position to do, then, is to grow as quickly as possible up to the soil surface so that it can begin producing its own food by photosynthesis. Until it reaches the light, why would it need leaves? It cannot begin photosynthesis until it reaches the light, anyway. So, no leaves of any size are produced, just tiny vestiges. The stored food in the seed might as well be used to extend the stem as quickly as possible through the soil to the light, not frittered away producing useless leaves or firm, sturdy stems in the dark of the soil.

As might be suspected, red light also will cause the leaves on etiolated seedlings to turn green as well as grow to normal size, and spindly stems to slow down their growth and become normal in appearance. And, again as might be suspected, infrared light reverses the effect of red light. Etiolated plants continue their rapid, spindly growth if a red light treatment is followed quickly by infrared light.

The important insight Borthwick and Hendricks drew from these observations on the effects of red and infrared light on seed germination, flowering, and etiolation was that surely they all must be linked to the same process. The similarities between the effects of red and infrared light suggested some factor common to them all. What could this factor be? Attempting to answer that question led these two investigators to make a brilliant and intuitive leap of the imagination.

What Borthwick and Hendricks suggested was that plants contained a single compound, a pigment (like chlorophyll in photosynthesis) that was capable of capturing (absorbing) both red and infrared light. Certainly both are a normal part of sunlight. Their idea was that this special pigment was converted to its active form by red light and back to its inactive form by infrared light. They also suggested that in darkness any of the pigment that had been previously converted to an active form by exposure to light reverted to an inactive form. Since there is more red than infrared light in the sun's spectrum, exposure of the pigment in the plant to normal sunlight would convert a large fraction of it to its active form. This conversion could then be used by the plant to sense whether or not it was in the presence of light.

Once having sensed the presence of light, a plant could then use the information as a trigger to set off a variety of different kinds of responses. In some species, the signal could be used to trigger the germination of seeds. In short- and long-day plants, the red light activation of the pigment could be used as a Zeitgeber to determine the length of day and night by sensing the appearance of light at dawn or its disappearance at dusk at different times during the seasons, a more complicated problem than simple seed germination but using the same triggering mechanism. In short- or long-day plants, this information could be used to sense whether a night was long enough or short enough, respectively, to trigger flowering. In the rapid growth of seedlings through soil, the red light activation could be used as the trigger to 'tell' the plant when daylight had been reached so that normal growth could begin.

These intuitive leaps arising from the speculations of Borthwick and Hendricks led, by the early 1960s, to the isolation from plants of a single blue-colored pigment (blue because it absorbs red and infrared light) that was given the name *phytochrome*. All flowering plants have it in all parts of their structure from the tips of stems to the tips of roots. Conifers, mosses, ferns, and at least some, if not all, algae also contain phytochrome or some pigment very like it.

The list of processes in plants under the control of phytochrome is growing all the time as we learn more about its effects. Of processes covered already in this chapter and the one before, certainly the germination of seeds, the growth of seedlings, and the formation of flowers are major events in the lives of plants under phytochrome influence. The resetting of the circadian clock by dim red light in the sleep movements of bean leaves as observed by Stoppel, Buenning, and Stern (see chapter 8) is also under phytochrome control.

The discovery of phytochrome was an absolutely crucial step in understanding how plants control their growth and development. The lives of plants are dictated by light. They exploit light as a source of energy to produce food in photosynthesis. Stem tips can sense the direction of light and use that information to orient shoots towards the light (see chapter 7). Now, we understand that information about the length of days and nights, and the seasons is also needed by plants so that their cycles of growth activity can unfold in an orderly way. For seed germination, etiolation, and flowering, the need for information about light and dark conditions is clear. To those needs should be added others such as the requirement for information about the approach of cooler, winter weather through the knowledge that days are becoming shorter and nights, longer. The puzzle of how it is that perennials produce winter buds long before temperatures begin to fall can then be solved. The plant 'knows' winter is not far away because the days are becoming shorter and the nights, longer, which it can sense with the aid of phytochrome. Thus, the plant can produce winter buds in mid-summer, or at any other time, if it wishes (and some trees, for example, do so in August in the northern hemisphere) with the information provided by red-light activation of phytochrome. Although I have emphasized so far the positive effects of light on plant growth and development, in some cases, a negative influence can also be used to advantage.

In desert species, exposure to light would indicate to a seed that it was at the surface of the ground and should *not* germinate because of the lack of water there or, perhaps, because of excessive heat. In these cases, the activation of phytochrome by red light when the seed is exposed at the surface of soil could be used as a signal not to germinate. Here is an example of the same trigger (red light activation of phytochrome) used in the opposite way, to inhibit seed germination.

Conversely, a seed buried deep underground could sense the *absence* of light, including red, through the absence of any active form of phytochrome. This information could, in turn, be used as a signal that the seed is in surroundings where it is safe to germinate since well below the soil surface moisture is more likely to be found.

Clearly, something like phytochrome would always be needed by organisms which spend their entire lives exposed to the elements, as all plants do. The close monitoring of day and night lengths along with the changing seasons is essential for survival of organisms which cannot protect themselves from the climate by moving to shelter as many other kinds of organisms do. Plants must have a way of sensing the approach of unfavorable seasons well in advance so they can make preparations to survive through to better times. That they also use the same device to accomplish many other essential tasks in their lives is no surprise. Recall from the previous chapter that the biological clock is used by humans to regulate a whole range of body functions from hormone release to urine production. Multiple use of a single reliable signal is not unusual in living things. Here, it has become obvious that a single signal (the active form of phytochrome) can be used to trigger many essential processes in the life of a plant. How might just one signal be linked to so many events at the same time? A simple analogy might help to illustrate how such links might work.

When an electrical switch on a wall is thrown an electric current is either switched on or off. What action then occurs depends on what the wire carrying the current is connected to. It may be linked to a light bulb, an electrical heater or fan, an alarm or any number of other appliances. The switch is just a trigger which can activate or deactivate any number of processes depending on the electrical connections.

So it is with phytochrome. In all cases, the 'switch' is the same. What switching phytochrome 'on' or 'off' triggers depends on what the switch is connected to. If the pigment is in a seed it may be connected to germination; if in a leafy branch, it may be linked to flowering; and if in an etiolated seedling, it may be a trigger connected to normal growth.

Phytochrome is such a vital molecule that the most lowly to the most sophisticated plants either have it or have something equivalent to it. Keeping in touch with the environment is vital to any organism. How to prosper rather than just survive in generally hostile, adverse environmental conditions is surely central to the highly successful life strategy of plants. Phytochrome is one key to how well a plant is able to carry out its life plan.

10

Dormancy: a matter of survival

In the previous chapter, we saw that for a successful lifestyle out in the open, exposed to the weather at all times, plants rely on an ability to bring the essential phases of their life cycles into line with changing seasons. The environments that many kinds of plants inhabit go through regular cycles of change in which seasons favorable for growth are separated by other periods when growth is either much slower or ceases entirely.

In preparation for the most unfavorable weather, a plant may need special protection against the climate. These periods of recurring poor growing conditions must be anticipated well in advance. It would be no use for the plant to begin preparing for a freezing winter, for example, on the morning of the first frost or for a long, dry, hot season in the desert at a time when water was no longer available. By then, beginning preparation would be quite futile.

What a plant has to do in preparation for long periods of poor weather is often quite elaborate and may take a considerable length of time in itself. There has to be a lengthy period of good weather after the signal is received by the plant that an unfavorable season is approaching. Thus, the signal received by the plant cannot be related directly to future poor conditions. For example, it is *not* low temperature which triggers the processes inside a plant leading to preparations for winter. The preparation might begin in mid- to late summer when the temperature outside is still high or even earlier in the growing season depending on the kind of plant and how its growth cycle operates. The shortening of daylength is more likely to be the advance signal plants use to anticipate approaching winter.

In most cases, advance preparations for adverse conditions are both elaborate and lengthy. In the formation of winter buds, the plant stops growing and, at the same time, stops producing new leaves from the buds at the ends of branches. Instead of continuing to form new leaves, small, tough scales are formed which tightly enclose the soft, delicate growing points inside. These scales are designed to take the abuses of winter weather without being destroyed. They can withstand snow, ice, and freezing and thawing many times over without disintegrating; and they repel water while keeping the tender tissues inside moist and alive. Only in the spring, when their task is complete, are they shed as the growing points begin once more to grow in length.

Many plants die back in the winter or during a dry season and retreat underground for protection from the climate. Underground stems or rhizomes are a favorite protective strategy in many such plants. Stem tubers are another similar strategy as are bulbs and corms. Stem tubers are distinct regions of the stem which become swollen while all the remaining parts of the plant progressively die back, potato tubers being a good, common example of this strategy. Bulbs (tulip), corms (hyacinth), swollen tap roots (carrots), and root tubers (dahlia) are all examples of the different ways plants achieve the same thing, the production of storage organs which can withstand long periods of adverse conditions. During the preceding growing season all these plants spend a lengthy period of time delivering food to these underground organs for storage in anticipation of the next growing season.

But above all of these strategies, important and successful as they are, we have the development of the seed as a means of survival during adverse climatic conditions. Seed plants have inherited the land. Two hundred million years ago, the earth was dominated by the 'spore bearers', plants such as the clubmosses and horsetails which flourished and grew often to enormous sizes, in great profusion. Some of the tree-like horsetails were 30 m high and had massive trunks rather like those of the modern palm. The ancient clubmosses, horsetails and another, now extinct, group, the calamites, dominated the earth's vegetation. But, as later with the dinosaurs, rapid climatic changes found these giants unable to adapt. Their massive corpses made rich fossil deposits in the shales, the origin of our coal deposits. They survive today as smaller plants often in the undercanopy of forests along with the ferns which also were spore bearers more dominant than they now are.

What is it about the seed bearers that has made them so much more successful in the more recent climates of the earth than spore bearing plants?

The development by some plants of the ability to form seeds was of very great importance. The seed is a product of sexual reproduction, the fertilization of an egg by pollen. Just as in animals, sexual reproduction is the basis of genetic diversity as well as being the bridge from one generation to the next. Many types of spores can be produced without sexual reproduction. Offspring produced when such spores germinate are identical to the parent. The plant produced from a seed is nearly always different in some way from its parents. Lack of genetic diversity puts spore bearers at a disadvantage during times of change in climatic conditions. And the climates across our planet have changed many times through the ages and will continue to do so. Genetic flexibility is a great advantage at times of change. If offspring resemble parents too closely and if the parents are ill-adapted to change, then their offspring more than likely will be also. Variability in offspring greatly increases the chances that at least some of them will survive, even if those few that still closely resemble their parents do not.

The seed is also a way for a plant to disperse itself in a highly protected way. The young plant embryo inside the seed is surrounded by a store of food to nurture it in the hours or days after germination begins. The tough outer shell or coat of the seed protects the embryo from adverse weather as well as from attack by disease organisms and predators. Dried seeds are just about indestructible. For example, many seeds, when just formed and still full of moisture, can be killed by exposure to temperatures around the freezing mark. The same seeds, when fully mature and dry, can withstand temperatures of −80 °C or lower. Durability is the hallmark of the seed, with few exceptions.

Many weed seeds can remain viable for decades although the majority of claims that some seeds have remained alive for centuries must at the very least be viewed as dubious. Claims that Indian lotus *(Nelumbo nucifera)* seeds apparently older than 1000 years have been germinated are questionable. The longest viability known is allegedly seeds of arctic lupine *(Lupinus arcticus)* discovered in permafrost conditions which can be dated to 10 000 or maybe even 15 000 years ago. Such claims should be viewed with caution. There are reports of the successful germination of wheat seeds from the ancient silos of Fayum stored for about 6500 years or from the tomb of Tutankhamen at Thebes (4000 to 5000 years old), for example, but detailed study of such seeds has shown them to be dead. There are a few credible estimates that put the ages of some viable seeds at over 600 years. Much more typical life spans for seeds are, however, from 10 to 50 years.

Not that seeds are always long-lived. Seeds of silver maple *(Acer saccharinum)*, wild rice *(Zizania aquatica)* and Japanese willow *(Salix japonica)*, for instance, lose viability in one week if kept in air. Others live for only a few weeks or months, dying if they lose even a small amount of their moisture or are exposed to lower than normal temperatures (seeds of tropical trees, for example). The seeds of major crop plants germinate within a very short time after planting, although the wild ancestors from which they originated may have various ways of delaying their seed germination. Domesticated varieties of such food plants have been bred over many centuries for quick germination. Farmers do not need, indeed do not want, any delay in the germination of their crop. They can determine for themselves when to plant or not plant a crop. A delay mechanism would be a great disadvantage, for instance in areas of the world where the growing season is very limited (northern North America, Europe, and Asia, for example) and often marginal in length for some of the crops grown there. Any delay, other than that imposed by the farmers, could mean no harvest before the onset of cold weather.

The characteristics displayed by most seeds are what make them of such great importance as food to humans. It is certainly no coincidence that the centers of development of the great human civilizations coincide with the regions where the major grain crops originated. Wheat, rice, maize and the pulses are outstanding examples of this. Seeds have proved to be by far our finest source of food supply in the most convenient form for storage.

But of interest to us here is how long-lived seeds remain viable for so long. We refer to this condition of longevity as *dormancy*. But what are the causes of dormancy? What advantages does dormancy have? How do seeds break out of their dormant state and germinate? What is dormancy, anyway?

When we use an ordinary word to describe a natural process we often do so rather loosely. In the present context, we tend to use the word dormancy or dormant in a very broad way to mean any state of suspended activity. Volcanoes are dormant between eruptions, as are hibernating animals between periods of activity. Insects are often said to be dormant before they emerge from their pupal stage. But in plants the word should be used with greater care. For example, plant growth can be slowed or even stopped for a time by lowering the temperature, although temperatures much below freezing will likely kill. The same plants when fully prepared for winter may be able to resist temperatures well below the freezing mark. So such plants are not regarded as being dormant just because lower temperatures slow or stop their growth. If freezing kills them they are not truly dormant.

True dormancy is a special property with its own set of conditions. Some people prefer to distinguish between true dormancy and what they call 'quiescence', which is just the prevention of growth by the absence of one of the basic conditions needed for growth, such as insufficient water or too low temperature. Supply the missing ingredient and the plant resumes growth. This is not so with true dormancy. Even under ideal growing conditions, dormant seeds, winter buds, or storage organs like bulbs and rhizomes usually will not resume growth at once. The state of suspended animation we call dormancy used by plants to avoid climatic stresses can often be very deep and not easily reversible.

The ability to avoid climatic stress has been of enormous importance in the evolution of plants, allowing them to colonize places that are hostile to growth for periods up to many years. Most plants have adopted a non-mobile lifestyle which is made possible because they manufacture their own food by photosynthesis from simple substances in the soil and the air. However, when it comes time to disperse themselves, they need to be mobile which means detaching a portion of themselves in a form which can withstand possibly long periods without water. This requirement is met by the production of seeds whose germination is under controls which are responsive to environmental conditions.

For a seed to germinate it needs water, oxygen, and the right temperature. Some germinate as soon as they are ripe but many seeds do not, even when conditions are ideal. This dormant period may simply be to ensure that a seed germinates in the spring following the year when it was produced. Spring usually provides better growth conditions than does fall when the following cold winter might destroy tender seedlings. Some Mediterranean plants are prevented from germinating by high temperatures, on the other hand, to protect seedlings from being subjected to scorching summer conditions. In many other cases, dormancy helps to ensure that at least some members of the batch of seeds produced by a plant in any one year germinate each succeeding year over a long period. In many plants, the clovers being an example, a proportion of seeds is programmed to germinate at once while others have a delay mechanism. The weed, fat hen *(Chenopodium album)*, is particularly elaborate having one kind of seed which is large and germinates immediately and three other, smaller kinds which germinate at various intervals. The fruit of cocklebur has two kinds of seeds, one dormant and the other, non-dormant. The non-dormant seed will germinate as soon as it leaves the plant; the dormant partner lies in the soil until the seed coat is damaged in some way.

One of the main characteristics of dormant organs is their ability to tolerate a low water content. As much as 90 percent of the weight of a growing plant is made up of water. In contrast, many seeds have 10 percent water or less, but not zero. Many seeds actually must dry out in this way before they are able to germinate successfully. This is an imposed dormancy mechanism typical of plants which grow in places with hot, dry seasons where seeds that germinate immediately would have great difficulty in establishing themselves.

In addition to dormancy based on the need for a seed to dry out before it will germinate, there are several other types each of which has its own characteristics.

One of the simplest and most effective types of dormancy is that imposed by the seed coat. Seeds made dormant in this way are often referred to by farmers as being 'hard' (clover is one good example). Water is prevented from entering hard seeds by a thick, waxy surface which has to be ruptured somehow before germination can begin. The breaching of the seed coat barrier can be achieved in many ways, such as by mechanical damage, insect attack, or invasion by fungi or bacteria. Each of these methods is subject to many variables so that some seeds may have their defenses breached quite quickly while others may lie in the soil for many months or years before water finds a way to penetrate them. In this way plants can spread out the germination of their seeds over long periods which increases the chances of at least some of them germinating successfully.

A grass or forest fire may weaken seed coats in large numbers and allow rapid and widespread germination among hard-seeded plants. In one case (an *Albizzia*, a small, legume tree found in Western Australia), the seed coat has in it a small corky plug such as in a wine bottle. The heat from a fire causes the plug to pop out thus breaching the seed coat and allowing water and oxygen then to enter. Seeds commonly found in conditions such as those of the chaparral of Mediterranean climates are very effectively breached by the fires that frequently engulf these areas. The frequent fires that sweep through southern California, fanned by the Santa Anna winds, are one good example. The result, though, is a rapid recovery of the area following a fire.

Soil movement due to ice formation and melting may cause abrasion of seed coats so that they become permeable to water. In farming, the coats of hard seeds are deliberately damaged just before they are planted so that water can enter at once. Passage through the digestive tracts of animals, notably birds, is another way to weaken a seed coat. In one bizarre example, a particular tomato which grows in the remote Galapagos Islands in the Pacific Ocean produces seeds which must pass through the digestive tract of the giant tortoise before they will germinate. Presumably, the slow, lengthy transit through the animal's gut is needed to complete the process of weakening this particular seed coat.

In other cases, the seed coat is no barrier to the entry of water and yet the seed still will not germinate for a period of time. In some cases the embryo can even be removed from the seed and nurtured carefully, yet it will not germinate. Some embryos, such as in the case of hogweed *(Heracleum sphondylium)*, are still not fully formed when the seed is shed and will not grow until their development is completed. In other cases, like the European ash tree *(Fraxinus excelsior)*, the embryos are complete but have not accumulated all the food reserves they need at the time the seed is shed. This type of dormancy is often associated with the need for a certain temperature range before germination will occur. For instance, hogweed embryo development takes place best at just 2 °C rather than at normal temperatures. The seed can then mature sometime over the winter and early spring, before germinating when the soil warms up the following year.

Light is essential to the germination of many seeds such as foxglove, tobacco, many primulas, and some lettuces. As we saw in the previous two chapters, light controls many stages in the life of a plant. It was from a study of the control of germination in lettuce seeds by light that led Borthwick and Hendricks to the discovery of phytochrome, the compound that all plants have and with which they receive the light stimulus in plant development (chapter 9).

Many plants show a kind of seed dormancy linked to light especially when the seeds are first formed. As they age in storage some seeds lose their responsiveness to light. Others, on the other hand, become more sensitive to light if attempts are made to germinate them at unfavorable temperatures. In fact, the responses to light are of every kind imaginable: sometimes germination is triggered by exposure to light; other seeds are inhibited by light; and others need daylengths of a certain kind, either long or short days, rather than just a simple exposure to light. The germination of foxgloves in woodland clearings is often spectacular, indicating that the seeds are capable of discerning not just the presence or absence of light but variations in light strength. This photocell-like capability in fact is used to measure light quality so that germination occurs only when the seeds are in bright, near full sunlight in woodland clearings, for example. In all cases, of course, the seeds are sensitive to light only when swollen with water and not when they are in a dry state.

Other seeds cannot bear light. Seeds of desert plants often have this property. In cases such as these, germination is usually triggered by moisture but only in the total absence of light. Such a response ensures that the seedling is buried and has some chance of rooting down before it is subjected to searing desert heat. Seeds of this kind can have their dormancy 'reset' by even a brief few minutes of exposure to light.

Many seeds, such as the stone fruits (including peach, cherry, and plum), some deciduous trees and several conifers, for instance, will not germinate until they have been exposed for weeks or even months to low temperatures and oxygen levels under moist conditions. Very rarely, the opposite (high temperature) is required and some will respond to alternating high and low temperatures. But so-called *prechilling* is by far the most usual form of temperature requirement for seed germination.

The need for a period of low temperature before germination will occur protects the seeds of plants found in temperate regions of the world from premature germination in the fall or during an unseasonally warm period in winter. In some cases, germination is merely speeded up; on the other hand many species have an absolute requirement for chilling and remain dormant unless exposed to temperatures close to 5 °C for a period of time.

A similar requirement occurs in the case of certain whole plants or plant organs (bulbs, storage roots, winter buds, etc.) which enter a state of winter dormancy that can only be broken after a period of prechilling. For example, if peach or apple embryos are removed from unchilled seeds and germinated separately, growth is very slow and abnormal. Stem elongation is slow and irregular; leaves are distorted. The symptoms can be relieved by chilling the stunted seedlings at 5 °C, a condition that in natural surroundings would be encountered sooner or later.

Desert temperatures, which are linked to the seasons, control the germination of local plants. Thus, in the Colorado desert of the USA, if it rains and the temperature is only 10 °C, it is mainly the winter annuals that germinate; between 26 and 30 °C, it is summer annuals that emerge. Cacti in such hot regions germinate best between 30 and 40 °C. Seedlings in these regions also sprout very rapidly to ensure their firm establishment before moisture runs out.

It is not difficult to understand failure to germinate in seeds where there is a mechanical barrier of some kind. Thus, in seeds with hard seed coats which have to be breached, there is a readily understandable, visible reason why germination can be delayed. The same is true of those seeds which contain immature embryos. Clearly, no germination is possible until the young plant inside the seed is completely formed. But what prevents germination in other cases for example, when seeds are complete yet will not germinate, sometimes for years, and in winter buds on deciduous trees or in storage organs like tubers, bulbs, and corms? What happens during prechilling? Why is light (or darkness) required? In many cases, seeds may be released inside succulent fruits and yet will not germinate. The tomato is a good example. Temperature inside a tomato is usually ideal for germination and there is ample moisture and oxygen, yet seeds will not germinate. Why not?

In the case of the tomato seed, the answer is fairly simple. The seeds will, in fact, germinate if removed from the fruit, indicating that there is nothing about the tomato seed itself which is lacking. It is just that the juice of the tomato fruit stops the germination process from happening. The seeds will germinate only after the fruit flesh has decayed or has been eaten by an animal. Seeds are then dispersed after traveling through the digestive system of the consumer. Many fleshy fruits inhibit the seeds inside them in the same way as tomato.

In many other cases of fruits, seeds, dormant buds, and storage organs, chemical inhibitors are present within the dormant organ itself which must either be leached out or destroyed in some way before germination will proceed. Sometimes, a particular growth substance block is present which arrests growth. Over a long period of time, the growth inhibitor is slowly destroyed and its influence over germination removed (see chapter 7).

Other chemical inhibitors present in seeds must be leached out before germination can proceed. In nature, enough rainfall to leach out these substances also leaves the ground wet enough to allow the survival of newly germinated seedlings. This is particularly important in deserts where lack of moisture is more of a limit on growth than any other factor.

In desert plants, brief showers will often have no influence on seed germination. Only relatively heavy rainfalls will provide enough moisture to leach out all the inhibiting chemicals and allow germination to begin.

Seeds certainly and maybe other kinds of dormant organs too, contain many kinds of inhibitors other than growth substances. The inhibitor may be as simple as table salt which, in the case of shadscale *(Atriplex confertifolia)*, is in such high concentration in the seed that it prevents germination until leached out. But usually the inhibitor is more complex than salt. Some seeds release cyanide, especially members of the rose family; others, ammonia. Yet others contain mustard oils, common in the brassicas, or alkaloids and steroids, and many other chemicals which we use ourselves as medicines. Plants have a wide range of chemicals which they use both to control their own development, including germination of their seeds, and, as we shall see in later chapters, to defend themselves against disease organisms, predators, and competition from other plants.

So, seasonal changes in the environment are responsible for controlling the cycles of growth and dormancy in plants, and for the timing of seed germination. The possession of mechanisms allowing environmental changes to be detected and timed is essential if plants are to survive in the majority of climates. But, while these responses must be efficient, they must also be sufficiently variable or adaptable to allow for unpredictable, unanticipated variations in the weather.

Thus to summarize, we can say that dormancy in seeds and other organs is one more component of the array of survival strategies plants employ. In some cases, we have seen that this strategy takes the form of a reduction in water content in seeds while they are still on the parent plant. Such seeds remain dormant until the completion of a minimum period of time has passed even if they are dispersed while there is still sufficient moisture available for germination. The seeds of winter annuals are good examples of this type of dormancy. When freshly produced, the seeds of many wild annuals will not germinate under normally favorable conditions, but only after being subjected to some very cool temperatures in their dried state. As the period of dryness progresses, they pass from a state of deep dormancy to one of relative dormancy during which they become able to germinate at gradually increasing temperature. Ultimately they reach a state where they will germinate at normal warm temperatures at which point dormancy can be said to be completely broken.

A difficulty arises in desert areas subject to infrequent and irregular rains. A light fall of rain might allow the germination of those seeds which have completed their dormancy period but might not be enough to support the growth of the plant produced. It has been shown that germination of such flower seeds occurs only after heavy rain even though light rain might have been sufficient to allow seeds to become swollen with moisture. It is thought that light rains are insufficient to wash out the inhibitors of germination lodged in seed coats. Continued rainfall for a period leaches out the inhibitors sufficiently to allow germination and also provides sufficient water to soak the soil so that plant growth, flowering, and seed set can proceed.

The requirement for chilling of seeds and buds is characteristic of plants inhabiting the so-called temperate zones, where adverse, cold winters are experienced. The requirement ensures that seeds and buds do not begin growth until the winter is past.

Light sensitivity is of value to dormant organs not just for sensing day or night in the environment but also for monitoring the quality of light. For example, a canopy of leaves overhead alters the wavelengths of light available by acting as a kind of filter. When light passes through the canopy, some of the red and blue wavelengths are absorbed by the leaves for photosynthesis. Since red light is required by some seeds for germination, the absence of red light can act as a signal to those seeds to remain dormant. Later, perhaps when the leaf canopy has thinned or has opened up for one reason or another, more red light will penetrate to the undercanopy and germination of these light-requiring seeds will take place. This may therefore be responsible for germination control in deciduous forests, or even under herbaceous plant cover, where many seeds germinate in early spring before the leaves have formed a canopy, and also for the large number of seedlings which appear in a forest clearing.

These and other mechanisms of dormancy emphasize the fact that the period between the formation of a seed and its germination is not just devoted to the problem of distribution and finding a suitable location in which to sprout. The period may be lengthened for other reasons associated with survival. In most cases these time lapses are concerned with ensuring germination either at some particularly propitious moment or as opportunities arise at intervals over longer periods of time.

In all cases, survival to the next season or to the next generation is 'the bottom line'.

11

Stressful tranquility

The last four chapters have focused on the normal life of a plant; how it grows, tells time, measures the seasons, and survives from one season or generation to the next.

But plants are also exposed to unusual, even extreme environmental conditions daily, not just from time to time or season by season. A landscape may appear to us to be restfully tranquil but, beneath the benign face of the placid green world, plants are waging battles constantly against the difficulties posed by their environments. Away from the equator, perennials, such as trees and shrubs, can be subjected to extreme cold in winter; plants growing at high altitude may experience, in addition to cold all year round (at least at night), drying winds and high levels of harmful ultraviolet radiation; desert plants often must suffer through long, difficult periods of extremely high temperatures; in many locations, extended periods of drought or flooding may have to be endured; a tolerance to saline soil may be a necessity; and more recently, soil, water, and air pollutants as a result of human activity must be added to the list of aggravations plants must cope with in their surroundings.

Because stress often leads to reduced health in plants, just as it does in animals, it is also of considerable interest to agricultural scientists. Stressed crops usually produce lower yields. Understanding how plants cope with and respond to stress is, therefore, important to the plant breeder whose job it is to develop crop varieties with resistance to natural stresses while maintaining high yields.

The word *stress* was used first by engineers when explaining what happens when a force is applied to an object. *Strain* is whatever change the object experiences because of the stress. For example, an elastic band can be stressed by using force to expand it. Strain, in this case, would be how much the band had been stretched by the force on it. As in this example, stress in the physical world can be precisely applied; the strain is often not so difficult to measure. But what about the biological world?

Measuring biological stress and strain is not so easy. For example, anything which does not allow a plant to reach its full potential (such as its growth, flowering, or seed production) might be thought of as a stress. Anything that, for instance, limits the ability of the plant to grow to its full size, form as many flowers as it is able, or set as much seed as it is capable might be viewed as being a stress leading to a consequent strain. Such a definition of stress and strain may be useful in agriculture where fairly precise estimates of the full potential of a crop can be made over several seasons by planting the same crop in different climates and different soils. Once this full potential is known, an estimate of how much a particular crop falls short of that ideal, because of drought, cold, or any other factor, can be calculated.

But what about a plant growing in the wild? How can full potential be calculated there? For example, many plants found in what seem to be the most stressful conditions on earth appear healthy and are often not found anywhere else. They seem happier where they are than in any other location. To take one of the most extreme examples to make the point: a few bacteria flourish in thermal springs at temperatures around 90 °C but they will not grow at 80 °C or at much above 100 °C. Can they be said to be stressed at 90 °C as they grow better at this scorching temperature than at any other?

There is no perfect way of defining stress in plants. For the sake of argument it is possible to say that stress is any outside influence which prevents a plant from functioning normally. Of course, 'normal' is, itself, highly subjective and varies from one species to another. Normal temperature in the case of the bacteria living in hot springs is around 90 °C. Then 80 °C would be a stressful, abnormal influence on organisms adapted to living at these even higher temperatures. Both of these temperatures, by the way, would cause the instant death of just about all other living things. So, both temperatures are definitely stressful and, yet, normal at the same time; it depends on the organism.

Of course, this same dilemma applies to any set of environmental conditions; what constitutes stress for one organism does not seem to be for another. For example, are the extreme environments found in deserts or on arctic tundra stressful for plants that normally thrive there? Normal, when it comes to the environment, is a moving target.

Whether or not we can define what exactly stress in plants is remains a topic of debate but at least we are beginning to understand some of the responses plants have to stresses placed upon them.

In the case of some plants, understanding reaction to stress is easy – they avoid it at all costs. These so-called *stress escapers* are often found in hot or cold (arctic) deserts where favorable conditions for growth are short-lived. Some plants which normally inhabit hot or cold deserts germinate, grow, and flower in a matter of days following seasonal rains or during short periods of summer warmth. Then, they form dormant seeds before the onset of unfavorable weather and never really expose themselves to stress. In fact, they have no special defenses against cold or drought since they rarely ever confront the extremes of the climate in which they live. They simply withdraw when the 'going gets tough'.

Many more plants, however, are able to resist environmental challenges through either *stress avoidance* or *stress tolerance*, which are two different things entirely.

Some plants attempt to reduce the impact of stress by avoiding it as much as possible without giving up entirely as the stress escapers do. Alfalfa, palms, and mesquite *(Prosopis glanduloso)*, for example, survive drought as adult plants by sending down very deep roots thereby making sure they have a secure water supply under conditions that other, shallow-rooted plants would find intolerable.

Cacti, with their fleshy stems and leaves reduced to mere thorns, are also stress avoiders. Their strategy is to take in as much water as possible when it is available and store it for the future. A cactus desert may look sparsely populated with lots of bare ground between each plant but just below ground level is a jungle of shallow roots spreading out in all directions in search of water. Even the slightest rainfall then provides some moisture which can immediately be sucked up by these surface roots and stored for future use.

The ability of succulents, including the cacti, to retain stored water is quite remarkable. In one celebrated case, a succulent, fleshy plant, a member of the cucumber family found in the Mexican desert, was stored, dry, in a museum for 8 years yet produced new growth every year although never watered. In the interval, the plant lost 4 kg in weight, mostly stored moisture used sparingly each growing season.

Plants do not have to be succulents to be efficient at conserving water. All kinds of strategies exist among plants to help cut down water loss. For instance, many kinds of desert plants have small leaves. Some may be thick and fleshy, and be used to store water as they are in cacti; others have small, flat leaves with low volume but large surface area by comparison, from which plenty of heat can be lost by convection air currents which circulate up and away from ground level into the atmosphere. This efficient heat loss helps lower the leaf temperature and reduces the need to cool the plant by other means, such as water evaporation. Significant savings in water can be made this way. Other plants reduce water loss either by clothing leaves in protective hairs, which may also be silvery in the shade to reflect light, or by shedding them during dry periods. The creosote plant *(Larrea divaricata* subspecies *tridenta)* is a good example of a desert species which simply sheds all its leaves and shuts down growth during the long, hot months or years of drought it must sometimes endure.

The hardiest higher plants are arctic and alpine species which often spend a good part of each year under snow. They may have to exist at low temperature much of their lives but are often somewhat insulated from the elements by snow cover. Many of these plants survive by adaptations to the way in which they grow. Some alpines stay small and low to the ground and are either covered in hairs or are coated in wax. They are often light colored, even to the extent of having grey-green leaves, to reflect excessive sunlight containing damaging ultraviolet light. Alpine and arctic grasses often grow in compact mounds so that dead and dying older leaves at the base of the mound form a protective mass within which young, tender shoots remain protected. In the center of these masses, the temperature may remain above freezing even when the air temperature round about is well below 0 °C.

Not that 'compact and small' is always the way plants protect themselves against the cold. High in the mountains of central Africa and in the Andes of South America, giant plants such as tree groundsels *(Senecio* species*)* and lobelias grow upwards of 6 m tall. Groundsels insulate themselves by having leaves covered in woolly hairs. Both groundsels and lobelias also have leaves arranged in such a way that they fold up every evening tightly around buds to protect them from freezing night temperatures. Some lobelias have another rather odd strategy for their protection. The leaves around buds are so tightly packed together that they can hold liquid as in a bowl. A watery fluid excreted by the plant collects in the bowl to a depth of several centimeters. At night, the fluid may freeze across but never to a depth of more than a centimeter or so. The main shoot tip remains submerged at the bottom of the bowl immersed in unfrozen fluid and protected from the worst of the cold.

Plants, indeed, have all manner of ingenious ways to avoid stresses. But, there are many others that equally boldly confront their environments by developing internal conditions to combat the stresses imposed on them. These are the so-called *stress tolerant* plants. Two types of tolerance exist; any one plant may have either or both kinds at the same time.

A plant species may be, what we call, *adapted* to tolerate a certain kind of stress. Plants with adaptations pass them on from one generation to the next by inheritance. The changes in the plant produced by the adaptation are usually obvious. They often involve changes to the structure of the plant and are quite visible, if not to the naked eye at least under a microscope.

Alternatively, a plant may become *acclimated* during its lifetime. Gradual exposure to a stress, such as cold temperatures or drying conditions, allows an individual plant to adjust its internal functions so that it can continue to live and reproduce normally. Of course, the *capacity* to adjust in this way must be inherited from previous generations. If there is no built-in capability to adjust to low temperatures, for example, then the plant will die if exposed to them. The capacity to respond to stress, then, is passed on from one generation to another. What form that response will take becomes obvious only if the plant is exposed to the kinds of stresses to which it and its ancestors have been acclimated.

In brief, we can say that acclimation is the changes a plant is capable of making to its functions *if* stressed *during its lifetime*. Adaptation is the permanent, inherited changes which have occurred to plant form and functions because of *long-term exposure* to a stress.

The adaptations plants make to their structure in different climates are well known and are a major part of what is called plant ecology. Already discussed here are the ways in which cacti and other succulents are adapted to their stark habitats in which they are, nonetheless, able to reduce moisture loss even under the scorching desert sun. Less obvious, but equally important, are the internal, nearly invisible tolerances plants have developed. Perhaps the most dramatic example of this discovered in recent times is the adaptation to climate found in the photosynthesis of tropical plants growing under hot, arid conditions.

As early as the 1920s, plants growing in temperate regions of the world were known not to be able to produce as much carbohydrate by photosynthesis at high temperatures as they were at low temperatures. Later, it was discovered that some of the carbon dioxide absorbed from the air was quickly being lost again to the atmosphere at high temperature even before it could be trapped by the action of photosynthesis inside the plant in the form of carbohydrate. When it was also found that this process involved not just the loss of carbon dioxide from the leaf but also the consumption of oxygen within the leaf, it was given the name *photo-respiration* (see also chapter 3). Fortunately, very high temperatures are infrequent in temperate zones of the world so that losses to carbohydrate production in leaves due to photorespiration are not so great in these regions. Not so in the tropics!

In the 1960s, first in sugarcane and then in maize, it was discovered that certain grass-like plants which had their origins in the tropics were different from the grasses typical of temperate climates. That they differed in their structure could clearly be seen even before other differences were discovered. Rings or haloes of darker, denser cells could be seen surrounding the veins when sugarcane and maize leaves were magnified under a microscope. Once closer attention was paid to this phenomenon, similar haloes were found in many other tropical and subtropical grasses, but not in grasses from temperate climates. Later, a special form of photosynthesis associated with certain grasses growing in tropical climates was linked to these features of leaf structure.

Tropical grasses like sugarcane and maize do not have photorespiration even at the high temperatures in which their wild relatives normally grow. These and other similar examples have a much more efficient way of delivering carbon dioxide from the air to where photosynthesis is occurring than do grasses found in temperate climates. Further, it was soon realized that this more efficient delivery of carbon dioxide was focused in the cells of haloes surrounding leaf veins. It is there that carbohydrate formation in photosynthesis is concentrated in these tropical grasses. Carbon dioxide captured in all other parts of any leaf is directed towards these special cells close to the veins. There, the carbon dioxide is trapped as carbohydrate with almost no loss back to the atmosphere.

The entire system of photosynthesis in these tropical grasses is designed for efficiency of: the capture of carbon dioxide by the entire leaf; the delivery of carbon dioxide in large amounts to the cells of the haloes close to the leaf veins; and movement of the carbohydrate formed by photosynthesis in the halo cells straight into the veins next door for transport to other parts of the plant (see also chapter 6).

The absence of photorespiration in these tropical grasses is an adaptation to the stresses found in the hot, dry places in which their ancestors evolved and survived. Seasons of drought coupled with heat impose on plants the need to conserve water. In those circumstances, a plant may be forced to close partially or completely the breathing pores (the stomata) in its leaves, that it uses for exchanging gases like oxygen and carbon dioxide with the atmosphere, simply in order to cut down water loss (see chapter 1).

But partially closing stomata will not just slow down water loss from the leaf but also the intake of the carbon dioxide needed for forming carbohydrate in photosynthesis. To compensate for restricting the size of their stomata in order to conserve water, the leaves of some tropical grasses have developed ways of efficiently sucking into their photosynthesis every molecule of carbon dioxide that comes their way; they fritter away none of it in wasteful processes like photorespiration.

We know now that other kinds of plants also have this special kind of adaptation, or variations on it, to hot dry conditions, not just tropical grasses (see chapter 2). At the other extreme are plants acclimated to life in freezing conditions.

Many plants, and animals too, can resist and survive the more or less severe freezing winter temperatures typical of the Arctic, subarctic and many temperate regions. Many insects, for example, living in these areas have the capacity to manufacture common glycerol when they need it to combat conditions below 0 °C. Glycerol acts as an antifreeze exactly as ethylene glycol does in water-cooled car engines. In higher plants, acclimation to freezing conditions is more complex than this but the outcome is the same.

Actual damage from cold is caused by ice forming between the cells of the plant. The problem with this is that water in solid form is useless to the plant which, in a sense then, suffers drought conditions inside its cells. Added to this later may be the collapse of the dehydrated cells under pressure from growing numbers of ice crystals just outside each cell wall. The water in these ice crystals comes from inside the cell leaving it dehydrated, deformed, and unable to function normally; the exact definition of a stress.

Frost-resistant plants have the capacity to defend themselves against low-temperature stress in a number of ways. Deep supercooling is common among those plants which inhabit climates where minimum temperatures in the winter reach –30 °C or lower. The boreal forests of Siberia and northern Canada, for example, are occupied largely by conifers that display this kind of acclimation to low-temperature stress. In deep supercooled tissues, water does not form ice crystals even at temperatures as low as –40 °C. At temperatures much lower than this, ice formation cannot be prevented and damage to the plant will occur. Although supercooling does not occur in all plants, it can frequently be found in the overwintering stems of large numbers of perennials in temperate zones and in many other plants that grow all the way up to the northern timberline.

Boreal deciduous trees and shrubs like paper birch *(Betula papyrifera)*, trembling aspen *(Populus tremuloides)* and willow, all of which may be found north of the Arctic Circle, survive because they can acclimate to the low temperatures of winter. During the normal growing season, all these plants will suffer injury or death if exposed even to mild frost. Yet, roots, shoots, buds, or other organs collected from these same plants after they have been acclimated to low temperatures can be stored in liquid nitrogen at −196 °C without injury! How can this be?

An answer to this question cannot yet be given in full but is being very seriously investigated, especially by scientists in the agricultural field. We would very much like to know precisely how plants acclimate to low temperature. Knowledge of how it is done could be of great significance to the development of frost-hardy crops. In some temperate regions of the world where the growing season is seriously limited by the number of frost-free days, knowledge of how to breed for frost hardiness is of great interest. All we can say here is that at least two steps seem to be involved in becoming acclimated, in woody plants, anyway.

First, in the early fall in temperate regions, when growth in perennial shrubs and trees is beginning to slow down but before leaves begin to fall, a signal is produced in the leaves and is sent to other parts of the plant. The production of this signal is triggered by short days, not by low temperature, and is linked to the action of phytochrome (see chapter 9). What form this signal takes and how it acts to protect the plant from the cold is still not known but probably involves the plant growth substance ABA.

The second stage of acclimation is linked to the first frost of the season. The advent of frost conditions is marked by many changes to the metabolism of the plant including an accumulation of several sugars and particular proteins called *dehydrins.* As temperatures fall, higher and higher levels of certain sugars accumulate in all parts of the plant. Most commonly, either glucose, fructose or sucrose, or combinations of the three, buildup in the plant sap. Less commonly, other substances like glycerol and other antifreezes accumulate in some plants. Dehydrin proteins protect cell contents against injury from dehydration. Together, all these substances serve either to keep the cell sap in liquid form, preventing it from turning to ice, or to protect against dehydration. As a consequence, fully acclimated woody perennial plant tissues are capable of withstanding temperatures far below those normally experienced in nature.

In annuals, such as herbs and crop plants, acclimation to low temperature also seems to involve ABA as well as the production of particular proteins that only appear in tissues where the cold stress is being felt. When exposed to low temperatures, the level of these proteins goes up in annuals, some in response to the temperature itself, others in response to the rise in ABA. Artificially adding ABA to these same plants before they are exposed to low-temperatures can protect them from freezing later and lead to the production of the same kinds of proteins as are formed naturally in response to low temperature stress. Some of these new proteins are dehydrins; the roles others play to protect the plant from freezing injury is still not clear. There is much to learn yet about frost hardiness in all kinds of plants including our most valuable crops.

Plants growing in hot deserts and semiarid areas of the world are constantly subjected to high temperature among other stresses. Flowering plants cannot survive temperatures much beyond 50 to 55 °C, although many cacti and *Agave* species can tolerate 60 °C routinely, while a few have been known to survive short exposures to temperatures around 75 °C. Even short exposure to such high temperatures induces most organisms to produce what have become known as *heat shock proteins* (HSPs).

HSPs are not just formed by plants in response to heat but have been found in a wide variety of animals and micro-organisms as well. And the amount of heat leading to the production of HSPs does not have to be much above normal. Exposures for a few minutes, in some cases, or a few hours, in others, to temperatures just 5 to 15 °C above the normal growing temperature are enough to trigger the formation of HSPs. They are produced with astonishing speed once their production is triggered. The first hints of their formation can be detected within 3 to 5 minutes of exposure to heat; within 30 minutes they are produced in the full amount; and they disappear once again within a few hours of relief from heat. So far, the exact roles of the HSPs are only sketchily known. Obviously, they somehow help to maintain the function of the tissues of the organisms put under stress but how they achieve this role is still not very clear. But their importance is without question.

As we learn more about HSPs, scientists are discovering that other, similar types of proteins also are formed when living things are put under stresses such as low temperature (already discussed), drought, when deprived of oxygen, or when grown in saline soils. It seems that all living things have ways of protecting themselves against a wide range of stresses in their environment. Some have developed very high levels of protection in order to survive in the harshest conditions. HSPs in plants probably provide just one glimpse of a range of special proteins designed to protect against the environment should it become hostile, either just from time to time or permanently. The roles of other kinds of protective proteins will undoubtedly come to light as investigations into how plants survive stresses continue.

Even then, plants find some stresses extremely difficult to deal with. One of these is the presence of salts.

One of the most serious stresses to which plants are subjected, especially from an agricultural point of view, is soil salinity. Of course, plants may encounter high concentrations of salts, especially sodium chloride, in coastal areas, like river estuaries and salt marshes, as well as in inland deserts. A limited number of plants have evolved to survive and even prosper in such conditions but the number of different species that have solved the problems created by salinity is small.

It is the increasing salinity of agricultural lands caused by heavy irrigation and overcropping that is, or should be, of increasing concern to us all. Water delivered to the soil by irrigation, especially, contains a mixture of salts which are left behind when the water evaporates. Overcropping causes depletion of soil nutrients which if not replaced leads to increasing soil salinity. Once certain levels of salinity are reached, affected land must be withdrawn from agricultural production since most important crop species are very sensitive to saline soil conditions. In China alone, for example, more than seven million hectares of agricultural land are classified as saline, much of this resulting from centuries of irrigation and overcropping.

Breeding of crops for resistance to salinity is a high priority in the worst affected parts of the world. Yet such an approach to the problem merely postpones the evil day since salts will just go on accumulating in more and more arable land unless the underlying problems associated with local agricultural practices are solved. Breeding for tolerance to saline conditions surely cannot be the long-term answer to a serious and growing agricultural problem.

Plants protect themselves against high levels of salts in three ways. Some like the mangrove, which grows with its roots in saline water, do not take in the salts; they have a way of simply not letting them enter their roots. Others take in excess salts but quickly export them again into salt glands on their leaves. In these cases, the salts crystallize on the surfaces of leaves where they can do no harm. Finally, some plants can take in quite large quantities of salts and then 'salt them away' in areas of the plant where they can do no harm.

But most plants are not very tolerant of salts. Of the important ones agriculturally some, such as beans, soybeans, rice, and maize, have almost no tolerance of excess salts at all. Tomatoes, cotton, sugarbeet, barley, and wheat will tolerate somewhat higher levels of salts but still not to any great degree. Barley is the most tolerant and can be grown, for example, on land made saline by irrigation on which other crops cannot be cultivated.

However, close examination of natural habitats where saline conditions are normal, like coastal salt marshes and inland deserts, suggests that salinity is a difficult stress for the great majority of plants to overcome. Only a few kinds of plants grow in areas where there is heavy salinity. Both sodium and chloride, the components of the most common salt, are toxic to plants in high concentration. Most plants avoid rather than confront this particular stress.

Finally, much is heard these days about environmental pollution – of soil, water, and air. Some pollutants, like smoke from forest fires started by lightning strikes, or gases and dust from volcanic eruptions, are natural in origin. Most, however, are the result of human activity. Pollutants in our air, waterways, and soil are virtually too numerous to count; the damage they are causing to ourselves, other animals and to plants, largely undetermined. The two most serious groups of pollutants with which plants have to cope are the heavy metals, found in soil and water, and toxic vapors in the atmosphere.

We saw in chapter 5 that all plants need certain minerals which they obtain from the soil. Unfortunately, though, increasing numbers of soils around the world also contain other minerals, such as heavy metals like cadmium, lead, and arsenic that are highly toxic, put there by human activity. Mining wastes, paper mill effluents, and deposits from emissions of gases into the atmosphere from heavy industries and automobiles, all are leading to increasing levels and more widespread distribution of heavy metals in the environment.

As with other stresses, plant species differ in their ability to tolerate pollutants. Some plants thrive on soils rich in arsenic, selenium, nickel, chromium, gold cyanide, cadmium, and other contaminants. As in the case of tolerance to salts, a few plants are capable of excluding heavy metals altogether, not even taking them into their roots. Others take up the pollutants in large quantities and accumulate them to levels that would be lethal to non-tolerant plants. These so-called *accumulator species* are sometimes useful in soil remediation, that is, in the clean-up of soil that has been contaminated by heavy metals. Polluted areas can be seeded with these accumulators which then take in the offending pollutants. Periodically, the plants can be harvested and the pollutants along with them. In this way, contaminants like selenium, nickel, zinc, and lead have been removed from polluted soils. The plants themselves are not harmed by the pollutants which are rendered harmless inside the plants. These very special plants protect themselves by binding harmful minerals to specially designed molecules they manufacture. These special binding molecules (which are called *phytochelatins*) are not produced by the plant until they are needed to carry out the job of neutralizing potentially harmful metals. In other words, some plants have the capacity to become acclimated to the presence of heavy metal pollutants in the soil.

Major gaseous pollutants found in the atmosphere that have an effect on plants are carbon monoxide, sulfur dioxide, various nitrogen oxides, fluorides, and ozone. Some of these are released during volcanic eruptions but most are produced from human sources such as car exhaust emissions, metal smelters, and coal-fired power plants. Most plants can handle and detoxify modest amounts of carbon, sulfur, and nitrogen gases and, to a limited extent, 'cleanse' the air around them. However, larger quantities of any one of these pollutants can cause major problems for many plants. Photosynthesis is particularly vulnerable to upset by these contaminants in the atmosphere.

Ozone, in particular it would appear, is now recognized as a major cause of injury to plants. The ozone layer high in the atmosphere may be of crucial benefit to us for filtering out ultraviolet light from the sun but ozone at lower levels in the atmosphere is a great danger, especially to plants.

Like other gases, ozone enters the plant through the stomata on leaves. Once inside the plant, the ozone quickly breaks down to produce a number of highly toxic compounds. These toxins are very aggressive and attack many natural, essential products inside the living cells rendering them useless to the plant. Photosynthesis is one of the most seriously affected processes. Lowering the efficiency of carbohydrate production of the plant by damaging its photosynthesis in turn lowers the yield of agricultural crops, the growth of forest trees, not to mention the destruction of flowers, shrubs, and trees growing along roadsides and in parks in urban areas. Excessive amounts of ozone in the atmosphere close to ground level is a major problem for many plants.

Although plants differ in their ability to withstand the stress of invasion by pollutants, as is the case for any other type of stress as we have seen in this chapter, mostly they have not had time to evolve the mechanisms needed to combat the greatly increased and increasing levels of these compounds. Pollution on the scale we now see it in some parts of the world was unknown even half a century ago. Plants, and animals too, have not been able to adapt or acclimate to such rapid changes. The consequences of this are all around us in our overpopulated cities and in the overcropped countryside. We ignore the signs at our peril.

Plants have evolved some remarkable and ingenious ways of dealing with environmental stresses. So far, their resilience in the face of variations in climates, natural and imposed by humans, has not been stretched beyond their ability to respond to and accommodate change. Increasing human activity, however, is increasing the severity of all the natural stresses faced by living things and is adding others not before encountered. Hamlet's description of the air as, 'but a foul and pestilent congregation of vapours', is becoming true in a way that Shakespeare could hardly be expected to have envisaged. Given that the resources of living things to withstand environmental stresses surely have some limit, the increasing burden of human activity must now be regarded as by far the most serious challenge faced by all life in the struggle for existence.

One ray of hope lies in the irony that animals are, ultimately, more dependent on plants for their continued well being than plants are on animals. It may be that plants will suffer decline as the burden of human excesses increases. But, as this burden on green plants escalates, so will the impact on human civilizations multiply. Green plants provide, through photosynthesis, both food to animals and replenishment of oxygen in the air they breathe. Thus, if photosynthesis were to decline significantly, so would the numbers of animals, including humans, eventually. But then, as human influence declined the plants remaining would have time to adapt to slower environmental change and would make a comeback.

Smart money is on plants to survive long after humans have brought about their own downfall!

12

The colorful world of plants

Plants produce a bewildering variety of exotic chemical compounds: the pigments which give color to leaves, petals, fruits, and seeds; the vast array of substances manufactured to create odors and aromas; and the equally large numbers formed to help defend against attack by disease organisms, predators, and competitors. Plants are complex chemical factories. We take advantage of the munificence of plants in many different ways, most notably, in medicine. World-wide, as we shall see in a later chapter, at least one-quarter of our medicines come directly from plants; much more than one-quarter in many societies. Most of these are compounds the plants use in their own defense against disease organisms and predators; fortuitously, they benefit us greatly, also. Many more compounds produced by plants have no known function although it may just be a matter of time before reasons for their manufacture are uncovered.

In the next five chapters, I want to pass on some highlights of what we know about a few of these interesting chemicals.

Plants and animals put a great deal of time and energy into attracting attention to themselves. By way of colors, scents, sounds, vibrations, and elaborate movements, such as those in mating displays, living things send signals to one another for many different reasons. The distances over which these signals are sent may be very short (a few centimeters) or long (kilometers in the case of some chemical sex attractants; the male gypsy moth can detect the scent put out by a female over 3 km away!).

Signalling is one of the essentials of life. Indeed, no plant or animal could afford the cost of giving off 'pointless' or random signals. The energy cost is just too great for any signal, whether it be color production, bird song, or perfume formation, to have no purpose. If there was no reason for these habits they would have been abandoned and eliminated from the living world long ago. Darwin in his earth-moving book, *The Origin of Species,* perhaps put it as well as anyone could in regard to flower color:

> *Flowers rank amongst the most beautiful productions of nature; but they have been rendered conspicuous in contrast with the green leaves, and in consequence at the same time beautiful, so that they may be easily observed by insects. I have come to this conclusion from finding it an invariable rule that when a flower is fertilized by the wind it never has a gaily-coloured corolla* [petals] . . . *if insects had not been developed on the face of the earth, our plants would not have been decked with beautiful flowers, but would have produced such poor flowers as we see in our fir, oak, nut and ash trees, on grasses, spinach, docks and nettles which are fertilized through the agency of the wind.*

The conclusions about flower color drawn by Darwin are very close to the opinions we still have today. Flowers are highly colored to attract mainly the insects which are such prominent agents of the pollination of flowering plants. Today, we would recognize, in addition, that some flowers are highly colored to attract other kinds of pollinators such as hummingbirds. Wind-pollinated flowers are usually inconspicuous since color is no advantage to them.

What Darwin said for flower color is just as valid for other forms of signalling. Of course, like any other activity, signals are fraught with dangerous side effects both for the sender of a signal and the prospective recipient. Thus, for example, while sending a signal to a prospective mate an individual may attract the unwelcome attention of a predator. In another instance, the berries of some plants are colored red or black to attract the attention of birds, the ultimate outcome being the dispersal of the seed in the fruit. Other berries may be red or black to warn various animals that they are poisonous or inedible. Sometimes, these signals can be confused by animals with unpleasant or fatal consequences.

We know how confusing color signals can be in the case of our own medicines. Pills may be red, yellow, or blue so that we can distinguish between our sleeping draughts, high blood pressure tablets, and cold capsules. Unfortunately, children may see the same colors in the sweets and candies they eat and mistake the pills for confections. The consequences of confusion in this case can be, from time to time, accidentally fatal.

So, we can accept that colors in flowers, and in fruits as we shall see later, serve useful purposes and that color is an essential ingredient in the plant world. What I want to do here is to outline some of what we know and understand about the purposes of color in the lives of plants and why there are so many different ones. Why do we see everything from nearly black to purple, blue, yellow, orange, red, and white in flowers and in fruits? How are so many colors achieved? Do plants make a whole range of different substances to achieve purple, orange, red, and all the other colors separately or are there just a few primary ones, as in the artist's palette, from which, by mixing, the full range can be formed?

A good place to begin a study of colors in plants is in fall leaves rather than by looking at flowers and fruits. The number of different colors in leaves in the fall of the year is less than in the case of flowers and fruits, and somewhat easier to appreciate as a starting point, therefore. As we shall see, though, the substances involved in producing color in leaves are the same as in flowers and fruits.

For those who live in temperate climates, at least, fall is marked by the advent of the rich, bright colors typical of the leaves of deciduous perennials just before they are shed. Many of the colors seen, but not all – the gold of birch, the flaming scarlet of sugar maple, the rich crimson of oak – have the same origin as the flowers of summer.

The simplest examples of fall colors are the yellows and oranges found in birch, elm, poplar, and many more of the trees common in temperate climates, as well as in most garden annuals towards the ends of their lives. These yellow and orange pigments (called carotenoids and found most conspicuously in the deeply orange-colored root of the carrot and in the mature pumpkin) are already present in leaves during the summer but are hidden from sight by the much larger amount of greenness, the photosynthetic pigment, chlorophyll. The greenness of chlorophyll (a word that means, literally, 'green leaf') is so strong that it generally masks the yellow and orange colors of the carotenoids in leaves. In the fall, however, the deciduous plant begins its annual, natural retrieval of useful materials from its leaves before it drops them; one result of this scavenging exercise is the breakdown of chlorophyll which, thus, disappears from leaves, allowing the yellow and orange pigments to show through.

The carotenoids are not just produced by leaves but are present in all parts of plants sometimes in large amounts. The annual production of these pigments by plants has been loosely estimated to be about 100 million tonnes, world-wide. Both animals and plants use small quantities of carotenoids as their source of vitamin A. For some reason, though, it seems necessary for many organisms to store much more of these pigments than can be justified by these and other, similar limited needs. For example, in a number of animals, considerable quantities of the carotenoids taken in with the foods they eat pass straight through the wall of the gut into fat and other body tissues. The golden color of cream or butter is one sign of these surplus carotenoids eaten and then stored by cows. In other cases, birds have all the carotenoids they need from the berries and seeds they eat to make their beaks, legs, or feathers red, yellow, or orange; in other words, birds use these pigments in their own signalling strategies.

The most spectacular colors of fall are the reds. Carotenoids can be red but most of the familiar red-colored fall leaves and many kinds of similarly colored flowers have in them another kind of pigment which is not often found in leaves during the summer. Only a few types of leaves, such as those of the copper beech *(Fagus sylvatica)* and some maize varieties, remain red throughout the summer season; a few varieties of ornamental plants are other notable exceptions. The reds of fall are members of a class of pigments quite different from those found in the same leaves during the summer. Why such striking amounts of these pigments should be produced in fall leaves is not known but they can be most strikingly seen in the maples as well as in trees of the pear family. They are best known to us in flowers rather than leaves, however; the first example from this group to be isolated from plants was actually blue, not red!

The first flower pigment to be studied was isolated from blue cornflowers (bachelor's-buttons) by the Frenchman, F. S. Morot, about 190 years ago. The blue pigment Morot identified was named 'anthocyanin' (the Greek word for 'blue flower'), a word now used for a whole family of similar chemical compounds ranging in color from blue through purple to red and pink.

The color of anthocyanins when they are inside the plant can differ from the color once they have been extracted. Thus, the red rose has the same pigment as the blue cornflower. The difference in color between the rose and the cornflower depends on what other substances are in the same place in the plant at the same time as the anthocyanins. The presence of these other substances in petals or leaves can change the color of the anthocyanin from purple to blue or red, for example.

Cool bright weather in early fall favors the development of brilliantly colored foliage. The brighter the light, the more anthocyanin is produced by those plants whose leaves normally turn red in the fall; the temperature must not be too high, either. In those areas of the world where fall is typified by mild and cloudy weather, fall foliage is duller, limited to yellows and browns; the red anthocyanins are not a prominent feature of fall foliage in these regions.

So, the pigments found in summer leaves are green chlorophyll and yellow and orange carotenoids. The carotenoids become more and more visible in the fall as chlorophyll begins to disappear from leaves towards the end of the growing season. In some plants, as chlorophyll fades away, a new family of pigments is produced by the leaf; red members of the anthocyanin family. Thus, the range of fall color is due mainly to only two families of pigments; the carotenoids and the anthocyanins, separately or together in different mixtures giving a range of subtly distinct hues.

In flower petals, these two families of pigments combine to produce an even wider range of coloration than is usual in fall leaves. The crimsons, purples, blues, and creamy whites as well as some reds and yellows are more often than not members of the anthocyanin family of pigments; the oranges as well as some other yellows and reds, are carotenoids. The greater variety of color in flowers is because the pigments bond with other compounds already in the petals. These other bonding compounds may be either simple minerals or more elaborate types of substances.

One example of the link between color and the bonding of pigments is the wide range of blueness in flowers. The kind of blueness found in some flowers is caused by anthocyanin bonding to one of the tannins commonly produced in plants. Tannins are not just found in places like tea (where we notice them, if at all, as brown stains on tea cups or teapots) but are common kinds of compounds in all plants.

But not all blue flowers are a combination of anthocyanins and tannins. Other combinations can give different tones of blueness. Thus, anthocyanin family compounds linked to the metal, iron, produce the blue color of cornflowers. In *Hydrangea,* if a certain balance between aluminum and molybdenum in petals exists, the flowers will be blue; if a different balance exists, the petals will be red.

In other cases, pigments combine with one another. The brown color of some petals, for example in the variegated wallflower, is due to a combination of a magenta anthocyanin and a yellow carotenoid. In other cases, variations in color in the same petals is due simply to the accumulation of the same pigments but in different amounts in different places. The deep purple coloration at the center of some poppies is the result of a higher local concentration of the same pigments that give the delicate mauve of the outer extremities of the petals of the same flower. These concentration differences can be quite large. For example, the anthocyanin in the 'normal' blue cornflower is present only in small quantities in petals. In the deep purple cornflower varieties, on the other hand, anthocyanin amounts can be 30 to 50 times higher than in normal varieties.

Finally, there are the pigments which produce the near white colors found in many flowers. These are really just members of the anthocyanin family which have no color to our eyes except, in some cases, a creaminess or an ivory tinge. They may not be anything spectacular to the human eye but these pigments are clearly visible to the eyes of bees and other insects that can see the ultraviolet range of light. None of the pigments is there strictly for our pleasure, we should remember; they all have a purpose. Whether *we* can see them or not is of no consequence to the plant.

Although there is a wide range of coloration in flowers the evolution of this coloration is by no means haphazard. There is mounting evidence of selection for particular colors in different environments, according to which are the most active pollinators present there. The preferences of the pollinators for certain colors can be crucial.

Much information is available about the color preferences of bees. They are known to prefer what appears to us as blue and yellow flowers. They can also discriminate among the types of anthocyanin pigments which to us appear white but which absorb strongly in the ultraviolet region of the spectrum of light. Although bees are not especially sensitive to red colors, they still visit some red-flowered species (poppies, for example) guided by other pigments in the same flowers which absorb light in the ultraviolet range but which the human eye cannot see.

The color preferences of other pollinators have been less well studied. Hummingbirds are especially sensitive to red; their preference for bright scarlet blooms, as in *Hibiscus,* is well known. Some other bird pollinators, such as sun-birds and honey-birds, appear to have similar color preferences to hummingbirds.

Other pollinators show less sensitivity to flower color. While butterflies are actively attracted to brightly colored blossoms, moths and wasps prefer duller, more drab colors. Finally, bats, flies, and beetles are not particularly attracted by color at all but depend mainly on other kinds of signals, such as aromas, to draw them to their host plants.

One of the most impressive color preferences is for fiery red among flowers which are pollinated by migrant hummingbirds. This is a fine example of signals which function at a distance rather than at close quarters. Quick recognition by migrating birds of the kinds of flowers that will provide them with food is critical. Since the birds pass rapidly from one area to another during migration they benefit from signals from their food sources which they can see from a distance. When they are in transit, it is not very practical for migrating birds to spend much effort on time-consuming foraging for food. Rapid recognition that possible food sources are close is especially critical to an animal like the hummingbird which, because of its large surface area compared to its tiny volume, needs a great deal of food to compensate for the excessive amount of energy it loses as heat.

In tropical habitats, where bird (particularly hummingbird) pollination is frequent, flowers are more often scarlet or orange in color. In temperate climates, more flowers are variations on blue, favored by bees and some other insects.

This is not to imply that blue flowers are absent from the tropics or red flowers from temperate regions. We know that this is not the case by any means. Bees and other insects do, after all, operate in the tropics and birds act as pollinators in temperate regions. What does seem to be true is that red and orange flowers are more frequent in the tropics and bluish flowers more prevalent in temperate climates. One very good example of this trend can be seen in a family of plants which has members both in the tropics and in more temperate regions.

Members of the Polemoniaceae (the phlox family) are common in the Americas, some in the more northerly, temperate regions of the continent and others in central, tropical areas. The hummingbird-pollinated examples of this family found in tropical latitudes have scarlet petals; bee-pollinated flowers found in the north are nearly all blue; and others are mauve or pink and are pollinated by butterflies. In the 18 species of this family, all the hummingbird-pollinated flowers have orange-red color, with only one exception; bee and bee–butterfly pollinated plants all have purple color; and butterfly-pollinated flowers are intermediate in hue, and contain various mixtures of orange-red and purple pigments. There is a very strong correlation between the type of pollinator (bird–bee–butterfly) and flower color in this single family of flowering plants.

Other examples among the families of flowering plants demonstrate that at least some species of plants can switch their flower color in just one or two generations to adapt to the availability of different kinds of pollinators. This has been observed in northern California, for example, where some kinds of plants growing out on open grassland and pollinated by bees have yellow flowers. Close by, in the darker redwood forests, the same kinds of plants are pollinated by moths and have less showy white or pale pink flowers. The latter colors are favored by moths which tend to pollinate at dusk or in the dark. Within one or two generations, plants that move from open grassland and relocate in the forest change the color of their flowers from yellow to white or pale pink; the reverse occurs if plants extend their range from the gloomier forest to more open territory.

For a plant to change the color of its flowers is not as difficult a trick to accomplish as might be thought. After all, as has already been pointed out, the majority of flower colorations involve just two groups of pigments, the anthocyanins and the carotenoids. By altering the relative amounts of members of these two families of pigments, different mixtures and, thereby, different colorations can be produced. Undoubtedly plant species have considerable flexibility in being able to stop, start, increase, or decrease the formation of these two families of pigments in flowers depending on what pollinators are available.

In addition to producing colors that are simple signals to particular kinds of pollinators, some plants have gone one step further and evolved ways of manipulating the behavior and movement of their favorite pollinators. Flowers in many families of plants do this by undergoing dramatic color changes as they age. These changing signals can be interpreted by pollinators to the benefit both of themselves and the plants they are targeting.

More than a century ago, Darwin arranged for the publication of a letter sent to him by the naturalist, Fritz Muller, in which Muller remarked on a multicolored *Lantana* growing in the Brazilian forest:

> *We have here a* Lantana *the flowers of which last three days, being yellow on the first, orange on the second, purple on the third. This plant is visited by various butterflies. As far as I have seen the purple flowers are never touched. Some species inserted their proboscis* [mouth parts] *both into yellow and orange flowers; others . . . exclusively into the yellow flowers of the first day. This is, I think, an interesting case. If the flowers fell off at the end of the first day the inflorescence* [flower] *would be much less conspicuous; if they did not change their color much time would be lost by the butterflies inserting their proboscis in already fertilized flowers.*

As Muller so intelligently observed floral color change benefits both plant and pollinator. The color phase of the flower provides an accurate indication of its sexual status and whether it still offers any nectar reward to a potential pollinator. In flowers where the color changes as the flower ages, not only is there a color difference before and after pollination, but a change in the amount of nectar as well. Plants that can change the color of their blossoms also tend to have abundant nectar with which to reward pollinators when their flowers are young. As the flowers age not only do they change color but they also contain much less nectar. Pollinators learn to link change in flower color to the availability of nectar. In doing this, they save the energy they would otherwise use in fruitlessly exploring flowers where there is no nectar or only small amounts of it.

One must be careful to say that the ability of particular pollinators to discriminate between flowers on the basis of color is not linked to the fact that they are the natural pollinators of those certain plants. In other words, this is not a case of co-evolution of plant and pollinator. New pollinators, which have never seen a certain type of plant before, when introduced to them, quickly learn which flowers offer reward in the form of nectar and which do not, using color as a cue. This recognition is a learned response, not one inherited by the pollinator.

We must also be careful not to conclude too emphatically that a particular color is the signal for a certain status. For example, red color in flowers does not invariably mean that there is a reward for a pollinator or even that the pollinator necessarily has to be a bird (because the flower is red). For example, one type of South American *Fuchsia* flower is green with purple streaks when it is young and contains abundant nectar. Later, when the flower is older, the color changes to red but these flowers have no nectar. The red-phase flowers in this case are ignored both by natural (bellbirds) and introduced (silvereyes and bumblebees) pollinators, whether birds or not.

In this case, then, redness is interpreted by the visitor as an indicator of the absence of nectar; the color change from green to red is used as an indicator of the end of nectar production. The green color indicating the presence of nectar lasts long enough for nearly 100 percent pollination to occur, but the color change takes place whether a flower has been pollinated or not. Color status is simply age-dependent and is just an indicator to a pollinator whether a particular flower is more or less likely to offer a nectar reward.

Current speculation about the redness of flowers and pollination by birds is that perhaps migratory birds use red as a long-distance signal. Plants may advertise their presence to birds passing through an area by this distinctive color which is so readily seen by birds. Such a strong signal may be less advantageous where only local birds are the pollinators. After all, birds have no programmed preference for red flowers, rather, preference is governed by the connection the bird learns to make between color and the nectar rewards available. The precise color is of only secondary importance.

Color, then, could help to advertise the presence of flowers and, often, their nectar reward status as well, at a distance. Once the pollinator is at close quarters, other signals (again color, but also often scents) take over and guide the visitor to the site of the reward. The so-called 'honey guides' are often part of the color patterning of flowers, their object being to guide the pollinator into the center of the flower, where the sex organs and nectar are located. They are particularly found in bee flowers and may be clearly visible to the human eye. A yellow spot on the lip of an otherwise blue flower or a series of dots on the petals leading the pollinator down to the area where the nectar, and incidentally the pollen and ovaries, can be found, act just like markings on a highway or motorway.

In some plants, honey guides are invisible to the human eye but can be seen in the ultraviolet range by insects. In black-eyed Susan (*Rudbeckia* species), in daylight, the petals are uniformly yellow to our eyes. If viewed in ultraviolet light, however, the outer parts of the petals are bright and the parts towards the center, dark. Members of the carotenoid family of pigments are responsible for the brightness of the outer parts of the petals in ultraviolet light and for the yellow coloration in daylight. The dark coloration of the central parts of the flower in ultraviolet light is caused by the presence there of members of the anthocyanin family of pigments.

Thus the yellow carotenoid pigmentation, visible in daylight, is used to attract pollinators from a distance. The anthocyanins at the center of the flower act as a honey guide once the pollinator has been lured in close by the carotenoids. This combination of carotenoid and anthocyanin family pigments is common in yellow flowers. Of course, other types of honey guides also exist.

Broadly speaking, then, Darwin was correct. Flowers are colored in order to attract pollinators; not just insects, as he supposed, but birds as well. Bats, flies, and beetles are also significant pollinators but rely more on the perfumes produced by flowers than their colors. Since Darwin's time, we have learned more about the roles colors play in providing signals to pollinators, not just at a distance but at close quarters as well. Of course, we also know much more about the kinds of pigments which form the colors as well as how they are produced. Another key role of color in plants is in the dispersal of seeds. As Darwin, again, observed in his book *The Origin of Species:*

> *that a ripe strawberry or cherry is pleasing to the eye as to the palate . . . will be admitted by everyone. But this beauty serves merely as a guide to birds and beasts, in order that the fruit may be devoured and the matured seeds disseminated. I infer that this is the case from having as yet found no exception to the rule that seeds are always thus disseminated when embedded within a fruit of any kind if it be coloured of any brilliant tint, or rendered conspicuous by being white or black.*

It is obviously desirable for a fruit not to be eaten until the seeds are ready for dispersal. The reds and purples of ripe fruit become so imprinted on the minds of dispersers that they will leave unripe fruit alone until the learned color signalling ripeness appears. Unripe fruits are often green, to camouflage them, bitter and sometimes spiny. The bitterness helps repel predators such as weevils and caterpillars. It is only when the seeds are ready for dispersal that fruits finally ripen and produce their brilliant reds, yellows, and blackish hues. Yet again coloration is the result of different mixtures of carotenoids and anthocyanins.

As fruits ripen, the sugars in them increase in amount to make the fruit more attractive to eat. Some produce subtle, 'fruity' scents which are useful in attracting seed dispersers such as bats, beetles, and monkeys. Birds either have a poor sense of smell or none at all but the fruit bats, especially, depend heavily on odor to guide them to their food targets.

Fruit-eating birds are the primary seed dispersers for plant species all over the world. Because seed dispersal is so important for plants, the criteria that birds use in choosing fruits should directly affect the reproductive success of plants that depend on bird dispersal of seeds. With their excellent visual acuity and color vision, birds undoubtedly use color as the major signal to locate and recognize ripe fruits just as in the case of flowers. Darwin noted that fruits eaten by birds are often brightly colored, at least to human eyes. Bright colors serve at least three functions: they draw the attention of a potential seed disperser; they make individual fruits stand out from surrounding green leaves; and they indicate ripeness.

A fruit's color may convey information about its quality that would influence a seed disperser's choice of the fruit. Obvious examples of such information are the changes in nutritional value and digestibility that a seed disperser can anticipate during ripening. But the disperser must also have a means to interpret the information, in context. If we take our own interpretation of fruit as an example. We know from experience that a red raspberry is ripe to eat. However, a red blackberry is not. Redness, in itself, is not a sign of ripeness to eat unless placed in context. A green apple may be unpalatable whereas a red or yellow one, ripe. However, a green pear may be fully ripe but a red or yellow cherry, still unripe. Nothing inherent in a color signals edibility or ripeness, just as in flowers it did not necessarily signal high nectar content. Color does, however, allow birds not just to identify a fruit but, at the same time, indirectly also receive information about its nutritional quality.

Color, then, serves at least two major functions in the lives of flowering plants. In both pollination and seed dispersal, the objective is the same: to attract suitable animal agents as efficiently as possible and at the most appropriate time. For us, at least, the bonus is the enhancement of our surroundings through the beauty of flower and fruit.

Finally, we so often assume that plants play a largely passive role in their relationships with animals. But, if we can learn one thing here, it is that plants are able, through a variety of signals, to influence directly the behavior and movement of the animals on which they depend. Color signals may be used to indicate to birds and other animals which fruits are ripe and ready for dispersal; floral nectar volumes, perhaps associated with flower color, can influence the length of time a pollinator will stay on a given plant thus increasing the chance of fertilization; and specific floral odors can attract the right pollinators at the moment of sexual maturity and receptiveness. Plants, in reality, play a remarkably *active* part in their interactions with animals. They are not just a passive, pretty backdrop in a landscape occupied by animals. They are consummate manipulators with dynamic strategies of their own.

13

Fragrance and flavor

Humans have used plant fragrances to enhance their attractiveness to one another since earliest times. For example, legend has it that Antony was attracted to Cleopatra as much by her irresistible perfumes as by her looks. Two of the three gifts brought to the infant Jesus by the Magi were frankincense and myrrh, both pungent, attractive plant resins. Frankincense is the gum produced by the olibanum tree *(Boswellia sacra)* and was used widely by the Egyptians, Greeks, and Romans as an ingredient of incense. Reddish-yellow in color, myrrh is produced by a number of the small, tropical bdellium shrubs *(Commiphora* species) native to Africa and southwest Asia. Myrrh is still used today in some cosmetics and perfumes as well as pharmaceuticals.

It is often not entirely clear why plants produce the odors they do. For example, many have special scent glands on leaf surfaces which are full of volatile oils and which burst, releasing their contents into the air, at the slightest touch. The purpose of most of these glands is not at all clear. Certainly animals can be very sensitive to leaf odors, as in the case of the well-known attraction of the domestic cat to catnip (or catmint). What advantage to the plant such lures may have is very often still a mystery. Other of these oils undoubtedly serve to repel herbivorous browsers and predators by making the leaf less palatable.

In the case of flowers, there is not much doubt, however. If a flower produces an odor, it usually does so to attract pollinators. The especially fragrant scents to human senses are most often those produced by flowers visited by bees and butterflies. The delightful perfumes of roses, lilacs, or gardenias are all there to attract insects. Odor is particularly important for night-flying pollinators. Bat- and moth-pollinated flowers, for example, often have very strong smells.

Not all odors are pleasant. Some, such as the skunk cabbage, are offensive, much like rotting meat. These odors, unpleasant to our senses, are, in fact, designed to attract certain kinds of pollinators. Carrion and dung insects, for example, may be attracted by smells released by some plants that often are very similar in composition to those given off by carrion or feces.

Of course, there are many other plant odors which are not involved in pollination. The sharp smell of cloves or cinnamon, the musky odor of thyme or sage, the fresh smell of a pine forest, the welcome odor of 'freshly percolated' coffee, the distinctive odors of lemons, oranges, and grapefruit, are all plant fragrances that are particularly familiar to us. Why plants produce these odors is sometimes a mystery, like those already mentioned released from leaf glands, since many seem to have no particular purpose. Those in fruits undoubtedly attract seed dispersers but many others have not yet been linked to specific purposes. Some of these, as we shall see later, have been exploited by humans particularly as spices and flavorings in a trade which is centuries old, was sometimes romantic, and at other times, vicious.

What is striking about all of these smells, whether produced by flowers to attract pollinators or by other plant parts, is that they are, on the one hand, so distinctive and yet at the same time usually so complex. In some cases, a single chemical may be produced by the plant which dominates the particular flower scent. More often, a mixture of many components makes up the fragrance. The smell of a rose is quite easily distinguished from lilac, yet, in both cases, the odor is the result of many chemicals in delicate balance with one another, not the product of just one substance. Some odors may be the result of a combination of dozens of compounds in different concentrations which, together, result in a certain, easily recognizable signature.

Why a flower smells the way it does is, therefore, often difficult to analyze. As we shall see later, blending to produce a certain smell is a large, important, and secretive aspect of the world's multi-billion-dollar perfume industry which so often attempts to mimic what plants achieve naturally. An important factor is that one component of a certain scent may reinforce a second and a third, producing a characteristic odor.

Chemists have attempted to analyze the nature of plant odors for many centuries. Most of the odoriferous plant compounds found in flowers are so-called *essential oils.* This is not a particularly good name for them because most plant perfumes are not oils at all as are those from, say, maize, olives or the castor plant. The name was given to them centuries ago, however, and such historical accidents are often difficult to undo. We are probably stuck forever with this misleading name.

The essential oils of plants are complex organic chemical compounds which will not dissolve in water but which, like their true oil and fat namesakes, are readily soluble in alcohol, ether, or chloroform. They are extremely volatile, that is, when released, they evaporate quickly into the air even at room temperature. The word *essence* in connection with the essential oils originates from the ease with which the smell of these substances spreads through the air.

There are close to 20 different chemical groupings which make up the essential oils. Some, like those found in pine trees, and spruce gum and resins, are grouped under the general name, *terpenes.* We have come across terpenes before but did not name them in this way. The orange and yellow carotenoid colors of flowers, fruits, and leaves (see chapter 12) are also terpenes. Natural rubber is, too, which illustrates how different from one another members of this group of chemicals can be. From colors in petals to some fragrances in flowers to gums like rubber seems a wide range, yet all are terpenes.

The various mixes of different terpenes produced by evergreen trees are so distinctive that we can tell one species from another simply by their smell. Other, less pleasant smells are also members of the terpene family. The smell of wet sheep's wool is the product of terpenes dissolved in the natural lanolin of the wool. The smell of fresh cut liver is the result of a terpene released by the damage caused by the knife.

Insects are very sensitive to tiny concentrations of chemicals in the air. As mentioned in the previous chapter, the male gypsy moth, for one example, can detect the sex scent put out by a female moth even at a distance of 3 km. It is likely, therefore, that flower odors are effective in attracting the likes of bees and butterflies at concentrations which, to our senses, are not very strong at all. Whether strongly or weakly scented to us, flowers synchronize the maximum output of their odors with the time when their pollen is mature and a flower ready for pollination. The output is even coordinated with the time of the day when the most favored pollinators are active. Thus, scent release is centered around midday for daytime pollinators and around dusk for pollinators active at night.

To us, flower scents seem to fall into two main categories. On the one hand, are the odors that to us are pleasantly perfumed; on the other, are those we would call decidedly unpleasant. The essential oils are generally the pleasing ones. The unpleasant odors have not been classified nearly as thoroughly as the perfumes, perhaps for obvious reasons. The perfume industry, for example, is not likely ever to be interested in a product based on the smell of rotting meat or dead fish, one can reasonably presume!

Unpleasant smells represent an attempt by a plant to mimic the odor of decaying protein or feces in order to deceive carrion and dung insects, which normally feed on dead or waste organic material, into transferring their attention to the flower and help pollinate it. The odors given off into the air are often extremely offensive and have names to match – cadaverine (as in cadaver) and putrescine (as in putrid) are two well-named examples. Others are unpleasantly fishy; yet others, rancid.

A few of the plants producing putrid smells have been investigated in detail and represent fascinating examples of this form of allurement. In the case of Jack-in-the-pulpit *(Arum maculatum)*, the blade-shaped spathe (the 'pulpit') that surrounds the central upright pillar, called the spadix ('Jack'), opens up during the night. At the same time, the spadix warms up to a temperature close to 30 °C. This burst of heat helps to evaporate quickly the objectionable odor put out by the spadix which, in turn, attracts dung beetles and flies.

When insects land on the smooth surface of the spadix they tend to slide down into the flower where they become trapped. The insects find it impossible to escape because the sides of the spathe and spadix are so smooth they provide no foothold. During the next 24 hours, in their frantic efforts to escape, the trapped insects transfer pollen and cause fertilization. Once this has been achieved, the flower begins to change very rapidly. One immediate alteration is in the surface of the spathe which becomes wrinkled; no longer smooth. Now, trapped insects can find some purchase, climb out of the flower and escape, their service to the plant fulfilled.

In another example among the carnivorous plants, the pitcher plant gives off a 'mousy' scent which is attractive to flies and other insects, although not for pollination. After enticing the insects into the pitcher trap, the same compound that produces the attractive odor also then paralyzes the victims, allowing then their leisurely digestion in the fluid at the bottom of the pitcher.

So, plants can deceive insects by producing attractive odors to trap them or to lead them to what they are fooled into recognizing as a source of food for themselves. In some instances, apparently, insects even go so far as to 'learn' to recognize the smells of individual types of plants. In so doing, these insects limit their attention to a small number or even only a single type of plant. There are even suggestions that a few plants give off hallucinogens and narcotics in their odors on which insects may become 'hooked'. Something approaching such a powerful attraction appears to operate in *Datura innoxia,* a relative of Jimson weed. Here, the nectar of the flower has dissolved in it some of the hallucinogens which are naturally produced by the plant. Direct observations have been made of the drunken flight of the hawkmoth after visiting these flowers at dusk.

Some flowers go so far as to mimic precisely the sex attractants of the insect pollinators they prefer. In one case, the sex attractant of the oriental fruit fly is given off in the scent of the golden shower tree. The release of this scent encourages the male of the species to visit the flower and, thence, help pollinate it.

In another example, one of the *Ophrys* orchids produces a flower which resembles in shape and color the female of a certain type of solitary bee. The male bee is attracted by the scent put out by the orchid flower which closely mimics the sexual odor of the female bee. In attempting to mate with the orchid flower, the male bee pollinates it. The attraction of the male to this particular flower is so powerful and so specific that the orchid has no need to produce any reward for the pollinator in the form of nectar; nor is the real female bee attracted to visit the flower, only the confused and deceived male.

The orchids, arguably, have the most sophisticated scent regimes of all the flowering plants for attracting and encouraging pollinators. In one further example, certain bees, living in the tropical forests of Central and South America, use the scents produced by orchids as sex attractants rather than wasting resources to form their own. As part of their sexual ritual, the males of these species congregate into swarms. Female bees are attracted to the swarms of males and mating occurs. While pollinating certain orchids, the male bees collect plant odors on their bodies and use them to attract males of the same species to form the swarms to which females are then lured.

Interestingly, different orchids attract different kinds of these swarming bees; in other words, although all the types of bees have the habit of swarming, not all kinds are attracted to all types of orchids. The orchid scents attract just certain kinds of bees, not others.

Plant odors, then, are distinctive, complex and, sometimes, aimed at very specific animal targets. In those associated with flowers, the principal aim is to attract pollinators, not to give pleasure (or in the case of foul odors, displeasure) to us. Attractiveness to seed dispersers, including humans and other animals, is no doubt part of the allure of fruit fragrances, however. Other odors, such as many produced in leaves and stems, have no known, precise function that has yet been discovered. Perhaps the exact targets for fragrant gums, resins, and many other pungent substances will become obvious in time.

What we can say is, it may be that plant odors are not designed for the exclusive pleasure of humans but that we have certainly not hesitated to make use of them and, in some notable instances, build from them multi-billion-dollar industries.

Essential oils have been put to use by humans for thousands of years. One of the earliest recorded means of using plant fragrances to scent the body was by burning materials like pine wood and other conifers, and then standing in the smoke. Indeed, our word 'perfume' comes from the Latin 'per', meaning 'through' and 'fumus', meaning 'smoke'. Burning scented herbs and inhaling the smoke was believed to have special powers. Linking medicine with religious ritual, the smoke would drive out evil spirits from the body and the illnesses or diseases associated with them. Modern aromatherapy is a link between the smoky origin of perfume and the healing qualities associated, down the centuries, with particular plant fragrances.

Cave altars in the Indus Valley show ceiling stains that provide evidence for the burning of woody incense to the gods. Baal Hammon, Lord of the Perfume Altar, a Phoenician god, is known to have been placated with fragrant smoke. The Israelites used incense to veil the presence of their God from human view. Even before this, the ancient Egyptians included aromatic wood shavings, gum resins, cloves, cinnamon, and tars in the wrappings of the dead to help preserve their bodies. To this day, the burning of incense is part of purifying rituals in many Christian churches.

An extensive trade in such fragrances as frankincense and myrrh developed around southern Arabia and the Red Sea over 2000 years ago. Because incense was an item of ritual it was not heavily taxed but the rarity of the plants in the area and the small amounts that could be harvested meant that it was expensive. The gifts of the Magi to the infant Jesus were, indeed, of royal quality.

But above all else, it was the development over 5000 years ago by the Egyptians of the art of perfumery that led to the main use of fragrances by humans. The ancient Egyptians used a wide variety of plant oils (olive, castor, and sesame) to anoint their hair and bodies, and perfumed the oils with fragrant herbs and gums. Cedarwood, juniper, cinnamon, frankincense, myrrh, and many other aromatic substances were blended to produce potent perfumes. By the time of the Greeks and then the Romans, not only were perfumes used lavishly by men as well as women on their bodies, but clothes, bedding, and the walls of their houses were anointed as well. At one time, Roman military governors thought it necessary to issue edicts prohibiting the sale of perfumes. It was thought that Roman soldiers were becoming effeminate and were being laughed at because of the attention they were paying to their *toilettes*. The entreaties of the authorities were largely ignored by a public enamored of their bodies. One ancient Greek poem set to music by Hubert Parry and entitled *A Lover's Garland* includes violets, narcissus, lilies, hyacinth, and roses as ingredients of the 'garland' which will soon adorn the 'scented locks' (and later, as the petals begin to fall, the bosom – 'O, happy they!') on the brow of the poet's lovely 'Heliodora'. In the face of such romantic obsession mere pedantic bureaucracy was powerless.

Naturally, the original, growing market for perfumes in Egypt led to an important trade in oils and gums between that country and Babylonia (present-day Iraq) and Arabia. Spectacular gardens of aromatics were cultivated in response to these consumer demands. A plantation of frankincense trees was developed on the banks of the Nile some 3500 years ago for Queen Hatshepsut; Cleopatra, the last Egyptian ruler (around 2035 years ago), owned a 'balsam garden' containing many plants and trees whose oils and resins perfumed her body. Jasmine has long been the origin of a favorite fragrant oil which is still in use today. Some have suggested that the word Gethsemane may be a mistranslation of 'Jessamine' and that the Garden of Gethsemane was a place where jasmine was grown. We shall probably never know for certain.

The collapse of the Roman Empire led to the end of the use of perfumes in Europe for a while. Nevertheless, the art of perfumery continued to flourish in such major centers of the day as the city of Alexandria. It was there, some 1400 years ago, that alchemists (the forebears of the chemists and pharmacists of today) invented a crude still for distilling alcohol. Soon after, it was discovered that the distinctive odors of different flowers could be dissolved in alcohol, giving us what became known as *essences* or *attars;* modern perfumery had been invented.

The most desired and most costly of the original pure essential oils was the attar of roses, notably still produced nowadays in the 'Valley of the Roses' near Sophia in Bulgaria. The attars of many other flowers quickly followed; jasmine, violet, and patchouli became great favorites down the centuries. Eau de Cologne from Germany, one of the most famous perfumed waters and still very popular, is a relative newcomer. It was invented just about 400 years ago and is a combination of rosemary, neroli (from orange blossom), and bergamot (a citrus tree somewhat like orange) oils distilled in grape alcohol.

With the discovery of musk as a preservative of essential oils, which have a tendency to turn rancid quickly if not stabilized in some way, a more reliable and widespread distribution of perfumes became possible. Since musk had been for centuries associated with male sexuality, the discovery that it would also act as a preservative for essential oils was good fortune indeed for the industry. The allure of perfumes which included musk was assured!

Alcoholic extracts of essential oils from flowers and other plant parts, stabilized by musk, made perfumes in the modern sense available to a Europe awakening after the Dark Ages. It was at this point that the French perfume industry had its beginning. Eleanor of Aquitaine, Queen to Louis VII of France (and later to Henry II of England) took particular interest in the revival of the use of perfumes in the royal courts. The French guild of perfumers was established about 800 years ago by Philip Augustus, Eleanor's son. The Crusades around this same time also brought back to France from the Middle East unfamiliar scents and the secrets of perfumery.

Yet, Italy, not France, became the first main center of perfume manufacture in Europe because of the favorable position of Venice as the main port of contact with the Middle East. But when Catherine de' Medici married Henri II of France in the sixteenth century, the transfer of the talent from Italy to France needed to build a perfume industry began. Certain areas of southern France were found to be ideal for growing the flowers required to provide essential oils and for producing the grapes from which the alcohol could be distilled that was needed for preparing essential oil attars. Soon, ways were found to incorporate essential oils into soaps, lotions, colognes, bath salts, and many other toilet products. There seemed no limit to the potential of the industry.

By the seventeenth century, sweet-smelling herb and nosegay gardens, begun in the monasteries of southern Europe and influenced by Roman villa gardens, had come to occupy an important place in many European countries. They supplied medicinal plants, and those for cooking, to the great town and country houses, as well as the ingredients for perfumes and attars. Less affluent citizens prepared sachets of dried flower petals which they hung around their necks or in their homes. Once again, as in Roman times, the overly extravagant use of scents caused laws to be passed in some places to restrict their use. It was ordained in some jurisdictions that any woman who lured a man into matrimony by the use of odors and 'other false objects' could be charged with procuring and the marriage annulled!

To this day, the center of the perfume industry remains in southern France. There, master blenders reign over an industry of great social and economic importance. A new perfume may take years to develop and millions of dollars to launch. Mainly for this reason, the highly complex formulae of ingredients of successful perfumes are highly guarded secrets; master blenders, highly valued.

Not all plant essential oils are used in perfumes, of course. Citronella is a powerful insect repellent while turpentine, extracted commercially from pine trees, is a major source of other chemicals. Spearmint, nutmeg, lavender, and lime can all be made from turpentine. One of the oldest and most famous of perfume ingredients, oil or attar of roses, continues to play a role in aromatherapy. It is one of the most antiseptic of the natural oils and one of the least toxic. Rose oil has long been valued for its soothing action on the nerves as well as for its powers as an antidepressant.

Whether using turpentine to produce the smell of fresh limes for commercial use in aftershave lotion, or rose petals to make simple scented water, plants leave us smelling and often, therefore, feeling a great deal better than we would be without them. Although we think of them primarily in sprays or stoppered bottles on dressing tables, perfumes also are essential components of hundreds of familiar household items. Besides their widespread use in toiletries and cosmetics, perfumes may well have helped us decide which writing paper, disinfectant, disposable facial tissue, or even floor polish to buy. Many of these nowadays are produced synthetically but all were, until recently, taken directly from natural sources, chiefly plants, suggested by their different aromas.

But there is another major and ancient industry also associated with the odors which plants produce. I refer to the spices and flavorings put to culinary use. Most spices also owe their valued properties to essential oils.

The prospect of the substantial amounts of money to be made in the spice trade lured adventurers from all over the world to the islands of the Indian Ocean over many centuries. Imperialistic governments fought over the Indian subcontinent, Malaysia, and southern China for centuries in order to corner and control the production and trade of spices. First the Arabians, then the Venetians, Portuguese, Dutch and, finally, the British came to dominate the East Indies where most of the spices originated. The ancient 'spices of Araby' – flavors, preservatives of meat, and luxuries of the rich – generated immense wealth for many of these colonialist countries. Medieval merchants and later Columbus sailed out to discover for themselves the rainforested areas of the world. Columbus, after all, was searching for a new route to access East Indian pepper, the only spice which could make decaying or heavily salted meat edible in his day, when he came upon the Americas, something of a disappointment at the time.

Eventually, however, spices of all kinds were smuggled away from where they had been grown originally and planted the world round thus breaking the Asian monopolies of centuries. In the twentieth century, such products as rubber and quinine from some of these same areas became of greater importance. Yet, spices from the tropics still earn at least 150 million dollars a year today in world trade.

Many of the essential oil flavors, such as vanilla and oil of lemon, are extracted in much the same way as perfumes and are used in their natural liquid form. Many other spices are made up of the parts of the plants themselves and are prepared by drying and grinding. Thus, cloves, peppers, ginger, mustards, sage, dill, hops, peppermint, spearmint, oregano, and many, many more have all contributed to varying degrees and in very many ways to what is still a vigorous trade, if not the dominant force it once was.

Although the chief uses of essential oils are for perfumery and flavoring, these do not by any means exhaust the ways in which essences serve humans.

Some of the powerful uses of essential oils and other plant products as medicinals will be dealt with in chapter 16. In addition, though, essences have proved most valuable in industry. One of the first plastics, celluloid, was manufactured from camphor. Turpentine has been used as a solvent and raw material for a multitude of products some of which were mentioned earlier in this chapter. Essential oils are incorporated in small quantities into manufactured products as diverse as adhesives, animal feeds, automobile finishing supplies, insecticides, and repellents, furniture polishes, janitorial supplies, paints, paper and printing inks, petroleum and chemical products, textile processing materials, veterinary supplies, and many others.

So, what probably began before recorded time as an attempt to render the body sweet-smelling and placate the gods through the generation of perfumed smoke has developed over the centuries into an integral part of a multitude of important industries. We may still not be sure why plants produce some of these aromatics but from Cleopatra onwards, we have learned to exploit them to our advantage in a multitude of different ways. Many, however, do play well documented, crucial roles in the lives of the plants that produce them.

14

Chemical warfare

In the wild, plants are surrounded by bacteria, fungi, nematodes, mites, insects, mammals, and other living hazards to their well-being, all of them hungry and many of them also potentially harmful. Plants cannot easily avoid these enemies by moving away or hiding. When you are fixed in one place by your roots running away from attack is not an option.

Plants can be devasted by attacks from one or other of their adversaries. Recall the near total damage to vegetation caused by plagues of locusts down the centuries or the wiping out of entire crops by disease as in the famous example of potato blight in Ireland in the mid-nineteenth century.

Yet, it is a fact that green plants still dominate the landscape despite the often proven ability of enemies to destroy them. Why is the countryside so dominated by plants and not by the other living things that are also present but less conspicuous? When we look at a landscape what we see, mainly, is green plants with relatively few animals scattered among them. Of course, if we were to look more closely, we would see more animal life than is apparent to the casual observer. Insects are much more numerous than appears to be the case at first glance, for example. Yet even taking into account the large numbers of smaller, less conspicuous animals, plants still make up something like 99 percent of the world's biomass, the total weight of all living things. Why is this so? Why do animals, insects in particular, not eat plants to near extinction or diseases devastate them to a far greater extent than they do?

Part of the answer is, of course, that plants have an amazing ability to renew themselves even as they are being attacked. Grazing animals, for example, may spend major segments of their days cropping the plants they prefer as food sources yet these same plants usually remain healthy and vigorous because of their ability to grow continuously as long as they have adequate light, moisture, soil conditions, and favorable temperatures. Drought conditions leading to overgrazing of weakened vegetation is a different matter. Disease may devastate a plant but rarely is the attack so complete as to wipe out the entire species. Renewal almost inevitably occurs, given enough time.

Another partial answer to the question of why plants dominate the world is that many of them have developed effective physical defenses. Some plants are too tough and leathery for most animals; others have deterrents such as rapier-like spines, thorns or prickles, or unpleasant stinging or tasting hairs on their surfaces making them altogether too formidable for most would-be predators to covet. Plants, in fact, seem to be generally unpalatable to other organisms and to have a wide variety of effective defenses. Only in selected cases does it seem that insects, grazing animals, and other organisms have the capability to overcome completely the wide array of defenses plants can muster against them.

Prominent within this formidable arsenal of weapons, we now appreciate more and more, is a bewildering number of chemicals plants produce to help combat attack. What are these compounds which seem to play such a key role in protecting plants? Some examples in this, and chapter 15, will provide an answer to this question.

Humans have understood for many centuries that plants can be hazardous to health. The lethal potion of hemlock (the alkaloid, coniine) given to the Greek philosopher, Socrates, was by no means the first occasion on which a plant product was used as a poison. But, the view we have of plant toxins is a rather narrow one. We tend to imagine that the only toxic plants are those which are dangerous to ourselves or to our domesticated animals. This is not the case. Many others can be highly dangerous to birds, fish, or insects, for example, even though they may seem to us innocuous. There are plant products such as nicotine, strychnine, and morphine, all of which are serious poisons to us, but we need to understand that, in addition, most plants are more or less toxic even though they do not contain particularly dangerous poisons. Eat too much of any wild plant and it is likely to leave you feeling at least mildly ill.

One critical point to remember is that the attack on plants by animals and microbes is nothing new but has been going on throughout the millions of years plants have been around. For all this time, therefore, plants have been faced with the task of protecting themselves, physically and chemically. So, in addition to producing the chemicals they use directly in their growth and development, such as sugars, proteins, and fats, plants also manufacture a formidable array of other substances, some of which they use in their own defense; many others of which still have no known purpose.

Chemists in the nineteenth and early twentieth centuries began the formidable task of identifying and cataloging a, seemingly, never-ending list of exotic chemicals which plants appeared to have but other kinds of organisms did not. For the longest time, most of these substances were thought to have no function; to be merely waste by-products. Some had been known for centuries for their importance to humans as perfumes and flavors (see chapter 13), drugs and poisons (see chapter 16), and industrial materials, but their value to the plants producing them was not so obvious.

Among these compounds, we now appreciate, are many examples of chemicals plants use in their defense. Presumably, as plants evolved over millions of years, they gained the ability to produce new kinds of chemicals from time to time. If some of these substances provided an advantage to the plant in some way (in warding off attack by predators, for example) then there would be a greater chance that the plant able to manufacture a certain chemical would survive to maturity and pass on its new-found capability to the next generation when breeding occurred. In this way, plants that had been better defended by producing that chemical would leave more descendants than undefended ones and be better able to pass on their new defense strategy to succeeding generations.

Not that predators then became mere helpless victims in the face of a new plant defense. As new chemicals appeared in plants, insect pests, herbivorous animals and disease microbes responded and over the generations developed new, inherited strategies of their own to combat the chemical obstacles put up by plants. In turn, plants once more responded by evolving yet more diabolical chemical defenses, and so the battle has continued to the present day. This is probably why some plants seem to have many chemicals which can be used to combat attacks by a variety of enemies and why others have fewer. The speed and direction of the development of defenses was likely dictated by the challenges faced by each particular type of plant, in its native environment, down the millenia. The greater the challenge, the more protection arose.

A consequence of their chemical defense strategy is that the compounds developed by plants to ward off attack by fungi, bacteria, and animal predators also make them less desirable as food for humans. The wild ancestors of our crops are often not nearly as palatable as their cultivated descendants. Our agricultural crops have been specially bred to lower or eliminate unpleasant compounds which may have been helpful in the wild. This is the reason many of our crop plants are more susceptible to insects and disease than are the wild ancestors from which they were developed.

Another point to remember is that the toxicity of any chemical is always relative. How harmful a substance is depends on the dose given and how quickly it is taken in. Even water can kill if drunk in sufficient amount over a short enough period of time. A dose of poison can be fatal if taken all at once but may not be lethal if imbibed a little at a time. The body may be able to handle something harmful if given enough time either to absorb it and deal with it effectively or begin passing it out again before the full, lethal dose builds up in the tissues.

A good example of this dose response to a poison is the case of the potato which has in it a harmful chemical called solanine. Fortunately, the amount of solanine present is so small that normally it is not a hazard in our diet. Potato tubers which grow above ground, however, and become green can have in them much higher quantities of the substance. 'Greened' potatoes can have sufficient solanine in them to cause victims to die of respiratory failure. The likelihood of death occurring depends on whether an individual has time to become accustomed to small amounts of the poison in the diet and can detoxify the substance as it is eaten.

Thus, plants have evolved many subtle ways to protect themselves through the formation of substances which for a long time were viewed as useless waste products arising from their everyday chemical activities. These defenses may be simple or elaborate. For example, some plants manufacture toxins that simply poison the attacker whereas others have less direct strategies, as we shall see later.

As already mentioned, insects and other predators have developed responses to the chemical warfare waged by plants. Many manage to adapt to plant defenses, at least to the limited extent that particular insects can overcome the defenses of certain plants, but by no means all. In this way, predators have developed feeding preferences; they will eat only the certain types of plants whose defenses they can handle, but not others to which they have no adaptation. Some insects have even progressed so far as to use the substances formed by plants for their own protection.

Thus, over millions of years of coexistence, plants and their antagonists have evolved in response to one another; each trying to neutralize the defensive or offensive strategy of the other in a never-ending competition.

There are three groups of plant products within which examples of defense compounds can be found. First are the terpenes, a group which also includes some plant growth substances (see chapter 7), as well as rubber, the essential oils which lend the characteristic odors to foliage (for example, peppermint, lemon, basil, and sage, see chapter 13), and the main red, orange, and yellow pigments found in flower petals (see chapter 12).

Many of the simplest terpenes are important in combating insect attack. One example is the pyrethrums which occur in the leaves and flowers of chrysanthemums. They are popular commercial insecticides because they are not very toxic to humans or their domesticated animals, and because they do not persist in the environment for very long and are, therefore, not serious contaminants of soil.

In conifers, such as pine and fir, simple terpenes accumulate in resin ducts in the needles, twigs, and trunk, and act as repellents to such insects as bark beetles, serious pests of conifers throughout the world. Others, in plants such as sunflower and sagebrush *(Artemisia tridentata)*, are located in the hairs on the surfaces of leaves and serve as repellents to anything that might have an urge to devour the leaf. To us, they taste very bitter. Yet others mimic the molting hormones of insects, making Peter Pans out of caterpillars, that is preventing them from ever maturing. If the caterpillars never 'grow up' to be adults, they also can never reproduce and any potential insect infestation is kept in check.

Perhaps the most powerful deterrent to insect feeding is azadirachtin a product of the neem tree *(Azadirachta indica)* of Africa and Asia. In doses as low as one part in ten million this potent plant terpene prevents insect feeding. For this reason alone it is being very actively and widely tested for its potential to limit insect damage to crops around the world.

In addition to terpenes, plants also produce two other diverse groups of chemicals that are used for a variety of purposes, including defense. One other group is that given the general name, *phenolics,* the most familiar example being, perhaps, salicylic acid from which we produce aspirin. Many simple phenolics have important roles in plants as defenses against insects and fungi.

Some of the most intriguing types of phenolics are those with an action only when exposed to light. Compounds of this type are found particularly often in foods like celery, parsnips, figs, and parsley. Phenolics can be transferred to the skin during handling of these foods and, in the presence of light, cause rashes. In insects, the consequence of eating these kinds of compounds can be much more serious than a mere rash; the insects may die if exposed to light after their meal. Some insects have adapted to surviving on plants containing these types of chemicals by living in rolled-up leaves; inside the rolled leaf, light is not bright enough to trigger the effect of these phenolics.

Next to cellulose, the most common substances in plants are the lignins, those phenolic compounds which give wood its great strength. The physical toughness of lignins deters feeding by animals. They are nearly impossible to digest and are, therefore, unpalatable to many herbivorous animals.

Another group of phenolic compounds which resemble the lignins are the tannins. The word tannin was first used to describe the compounds in plants that would convert raw animal hides into leather by the process we call 'tanning'. In fact, tannins are very widespread toxins which greatly reduce the growth and survival of many herbivorous animals when added to their diets. In humans, tannins cause a sharp, unpleasant, astringent sensation in the mouth. Unripe fruits, red wine, and well-brewed tea all are abundant in tannins. In animals other than ourselves they cause loss of appetite; in many microbes, they simply stop growth.

The last group of plant products active in defense are certain complex molecules all of which have at least some nitrogen in them. Familiar to us are alkaloids such as nicotine, codeine, morphine, cocaine, and strychnine, one or more of which are present in about one-third of flowering plants. All alkaloids when taken in high enough dose are toxic and act to defend the organisms containing them; at lower doses we may find them useful as medical or non-medical agents, and as stimulants or, in contrast, as sedatives (see chapter 16).

Various of these nitrogen-containing compounds in addition to alkaloids also play a role in the defenses of many plants. Several give off cyanide when a plant is crushed. Normally, such potentially dangerous substances are stored safely inside the plant but when an insect, slug, snail, or other herbivorous animal damages the plant tissues, these compounds are released and produce cyanide gas which can cause the death of or, at the very least, illness in the attacker. Cyanide defense is quite common, being found in plants as diverse as clover, Sorghum, Lotus, bracken, and the desert plant, jojoba *(Simmondsia chinensis)*.

Mustard oils also produce volatile defense gases when released during attacks by predators. The mustard oils are responsible for the smell and taste of such vegetables as cabbage, Brussels sprouts, broccoli, and radishes. They act as toxins or at least as very good repellents as anyone knows who has tried to tempt young children to eat any of these vegetables!

Two main strategies are employed by plants in using toxic compounds to defend themselves against enemies. One is the simple accumulation of a toxin, or a mixture of toxins, from the earliest stages of development of the seedling throughout the life of the plant. In chicory, for example, three defense compounds are produced which are extremely bitter and which deter all animals, including insects. Their effectiveness is increased by having them all included in a sticky latex that oozes out of the tissues in any part of the plant when damage occurs. There are some other obvious ones, like deadly nightshade *(Atropa belladonna)* and hellebore; yet others less well known, like *Laburnum*, lupines, *Delphinium* and groundsel; and some surprises among our staple vegetables.

I have already mentioned the toxicity of the solanine in potato but there is also cassava, a tropical root, which has to be specially treated to remove the cyanide it contains. Cyanide also occurs in almonds, lima beans and apple pips, sometimes in quite alarming amounts. In fact, in some cases of vegetables, if, by law, the harmful natural compounds they contain had to be listed on their packages, we would be forced to remove them from supermarket shelves in short order! Imagine artificially *adding* a sublethal dose of cyanide to a food product, as in cassava, and then attempting to have it approved for human consumption! Then there is the oxalic acid in spinach and in the leaves of rhubarb. Fortunately, the amount in spinach is quite moderate but in rhubarb leaves it is there in dangerous quantities – a very good reason not to eat them.

All of these, and there are many more, are examples of static types of defenses; there all the time in high enough quantity to act as deterrents. But to maintain such high levels of quite complex chemicals in all parts of the plant at all times is very costly to the plant. There is a significant cost in producing these substances; a cost in energy and materials that could be used for other purposes.

To moderate the cost, some plants have developed the strategy of restricting toxin production to tissues most vulnerable to attack. For instance in the coffee plant, caffeine is formed in young, soft, juicy leaves that are attractive to predators to a level of about 4 percent of the dry weight of the leaf. As a leaf ages it thickens and toughens, and becomes less attractive for eating. The rate of caffeine production decreases to just a fraction of a percent. Later, when the soft, young coffee bean has just been formed, the concentration of caffeine in the bean increases once more to about 2 percent of the dry weight before declining once again as the bean hardens and ages.

An even more dramatic example of chemical protection being concentrated in vulnerable plant organs occurs in the paper birch, a native of

Alaska. Here, a certain member of the terpene family is accumulated to a massive extent in the young, succulent parts of the stems and twigs, portions of the plant favored as food by the snowshoe hare. The concentration of the deterrent is 25 times greater in these young tissues than in the more mature parts of the tree and acts as a very effective deterrent against grazing animals, notably the hare.

Another form of protection used by plants is to have different chemicals in different parts of the plant at different times within a growing season. Thus, some plants may have one chemical defense in the first leaves produced in a season and another in leaves formed later in the year. Thus, leaves lower on the stems and branches of some plants produced early in a growing season contain different deterrents from younger leaves produced later, higher up the plant.

Some forms of protection employed by plants are more subtle and sophisticated than just the production of unpleasant or lethal toxins. First among these strategists are those plants which produce and then accumulate massive amounts of insect hormones. Numerous examples have now been recorded of plants, especially conifers and ferns, that contain compounds which, when eaten by insects, upset the normal steps in their reproduction leading to sterility and death. A few years ago massive amounts of an insect hormone were isolated from the leaves of one of the sedges; enough of the compound to cause sterility in a wide range of insect predators, including grasshoppers.

Some plants can mimic many substances other than hormones that are naturally formed by insects. For example, an aphid which normally preys on such plants as the wild potato secretes a substance called an 'alarm pheromone' when it is attacked by a predator. This gaseous substance alerts other aphids close by that they, too, may be in danger and should escape. The same substance is found in the hairs on the surfaces of wild potato leaves. When attacked by aphids, the pheromone is released from the leaf hairs; in turn, other aphids in the vicinity are driven off. The plant repels a major pest by mimicking the pest's alarm signal, a near perfect deterrent. After all, an insect could not readily counter or adapt to this type of defense since it surely could never ignore the very chemical signal that is so critical to its survival.

The leaves of long-lived plants are particularly vulnerable to attacks by predators. In the case of a tree, juicy leaves are produced year after year, maybe for centuries, time enough for insect populations to adapt to and overcome defense strategies adopted by plants of this kind. Obviously, determined insect predation could devastate a tree over time, something which can and does happen occasionally. Why does it not happen more often?

Patterns of insects' grazing on trees show that only some of the leaves on any one tree receive close attention from these predators. Other leaves on the same tree are not attacked to nearly the same extent. On closer analysis, it is often found that the leaves most affected have much lower levels of defense compounds than those less frequently grazed. Levels of tannins, for example, have been found to vary from leaf to leaf even on the same branch. Yet, one leaf is indistinguishable from another; there is no way of telling which leaf is palatable until a sample is taken. Therefore, all leaves have to be tested so as to discover which are palatable. This not only wastes the insects' energy but also makes it more vulnerable to predators such as birds. In moving around in search of leaves fit to eat, an insect becomes more visible to a watching enemy. Indirectly, forcing individuals to act in this way aids the plant by eliminating individual members of the insect population attacking it.

This same kind of variable defense seems to work not just within a single plant but also between plants. In the tropics, there are examples of tree communities in which one member may have low levels of protective tannins while the tree next to it may have much more deterrent chemical. An insect predator still has an unpredictable food source and must spend more time moving from tree to tree searching out low-tannin individuals.

There is usually a heavy cost in materials and energy in providing protection. Thus, the incidence of variable deterrence is quite widespread in the plant kingdom. In the case of the tropical trees in the previous example, high-tannin individuals compensate for the energy and material used in providing for their own protection by forming fewer leaves than their low-tannin neighbors. It has been estimated that if half the leaves on a plant or half the individuals in a community of plants produce protective chemicals, there is close to 100 percent protection of the whole community. In other words, if an insect is faced with a population of plants, half of which are unpalatable, but it does not know which half, then it is not worth the insect's time and energy to seek out the unprotected individuals. Better to move on and find less well protected food sources.

So far, all of the examples given have been of plants that produce high levels of protective chemicals, although not necessarily in all individuals, as we have just seen, or at all stages of their lives. This is a costly strategy which constantly uses up valuable materials and energy that could be used to greater advantage in other ways. A better strategy would be to produce the protective compounds only when they are needed, that is, as an attack is taking place.

Induced defense was first recognized only about a quarter of a century ago but has now been recorded in a wide variety of plants. In trees, this type of defense strategy usually involves the quick production of tannins, in response to an insect infestation for example. In herbs, the toxin is more likely to be an alkaloid or a terpene of some kind.

One case in point occurs in an attack by the hornworm on wild tobacco where alkaloid production reaches a maximum about 10 days after invasion begins. Nicotine and nornicotine increase by over 200 percent. Interestingly, these alkaloids are not produced in the leaves where the attack is taking place but in the roots from where they are moved to the leaves as the attack continues. How do the roots 'know' that an attack is taking place in the leaves and should begin nicotine production? This is a question that has occupied scientists for many years; the way in which plant tissues send signals to one another.

Caterpillars grazing on tomato leaves, for example, do not go unnoticed. The unwounded leaves of a plant under attack somehow receive an alarm signal from a wounded neighbor which causes them to begin preparing to be attacked themselves. Long before they are actually invaded, untouched leaves begin manufacturing defense compounds. The identity of the agents which travel rapidly from the invaded to the remaining leaves and raise the alarm has been sought for many years. Likely candidates have come and gone over the years but the debate is likely to go on for a long time yet.

One of the more recent ideas about plant alarm signals is that, in tomato, the damage caused by an insect initiates an electrical current which travels out to adjacent, untouched leaves and triggers the production of jasmonic acid which, in turn, initiates the formation of defense compounds which, when eaten, prevent digestion of food by the predator. Whatever the initial alarm signal eventually turns out to be, chemical or electrical, that notice of attack is sent from one leaf to another in the tomato, and some other plants, there is no doubt.

More controversial is the idea of an alarm response from one *plant* to another. Studies with the Sitka willow *(Salix sitchensis)* suggest that attacks by insects not only cause the leaves on the same tree to change and become less palatable to the attacker but that leaves of willow trees nearby also change. How this could occur remains a mystery. One suggestion is that damaged leaves give off a gaseous signal which travels through the air and alerts trees of the same kind nearby. What is not so clear is whether the signal is, indeed, gaseous and can travel by air or whether, for example, the roots of adjacent trees are touching so that some substance could travel between the trees underground. Another possibility is that a few insects which have just fed on one tree carry some chemical to the next target and release it into the new host thus inducing a rapid defense buildup. Whatever the final answer to these questions is, the idea that plants may carry out airborne chemical communication is intriguing.

Some insects have been able to adapt to the chemical defenses of plants and in some cases even use them to their own advantage. One example of the latter ability is a grasshopper that feeds solely on plants of the milkweeds *(Asclepias* species). The milkweeds manufacture a number of complex compounds that can severely disrupt normal heart function. The grasshopper defends itself from attack by other predators by spraying a poison from a gland. Upon analysis, the spray was shown to contain the very same toxins as were produced by the milkweed. If the grasshoppers are maintained on a diet free from milkweeds, the poison content of the insect spray falls dramatically. There seems no doubt that the source of the poison used by the insect is the milkweeds on which it normally feeds.

Another very well known example of this same plant-to-insect defense is the case of the monarch butterfly and, again, in association with milkweeds. In the 1960s, in Florida, it was noticed that monarchs reared from eggs on milkweed plants were not acceptable as food to the bird predator, the blue jay. Birds that ate the butterflies became violently sick. Again, analysis showed that the bodies of the butterflies contained enough of the toxins normally produced by milkweeds to kill an animal as large as a cat or a dog several times over. The monarchs themselves were completely immune to the toxins. Furthermore, butterflies reared on milkweed toxin-free diet had no effect on blue jays.

This classic milkweed/monarch butterfly/blue jay case has been studied intensively since its discovery, investigations that have led to other discoveries about the significant influence plant defense chemicals can have in the animal world.

One consequence is that other kinds of butterflies protect themselves against bird attack by mimicking the coloration of the monarch. These mimics, such as the viceroy butterfly, are not poisonous but because they are look-alikes, they are, so to speak, under the umbrella of protection of the monarch while avoiding the need to store the milkweed toxic chemicals or develop their own immunity to them.

One more interesting point associated with monarch butterflies arises from observations made at their overwintering ground in central Mexico. Here, they are the victims of black-backed orioles and black-headed grosbeaks which together account for 60 percent of monarch deaths. These two kinds of birds, however, feed very differently on their prey. The orioles, which are sensitive to the toxins packed by the monarchs, pick the butterflies apart carefully, selectively removing certain muscles and parts of the body where no toxin is found. Grosbeaks, on the other hand, are much less sensitive to the toxins and feed wholesale on monarch carcasses.

The case of the monarch butterfly is a good illustration of the fact that, no matter how good a defense is, none can give complete protection, all of the time.

One of the most striking examples of an insect adapting to the point where a plant's protective chemical has become essential to the survival of its predator is the case of the beetle, *Caryedes brasiliensis*, which is the sole predator of the seeds of the vine-like legume, *Dioclea magacarpa*, that grows in the deciduous forests of Costa Rica. About 13 percent of the seed dry matter is made up of L-canavanine which, although it is an amino acid, is not one of the 20 usually used to form proteins (see chapter 4). L-canavanine is also a potent insecticide. This particular beetle, however, uses the L-canavanine in the legume seed as its main source of nitrogen as it progresses through its larval stages. The beetle has evolved a way of forming urea from the L-canavanine and, then, converting the urea into ammonia which it then uses, in turn, to form nearly every amino acid it needs to produce the proteins in its larval body. Thus, the potentially poisonous defense compound of the legume has become an essential ingredient for a beetle predator.

The classic, older view, that the many exotic chemicals produced by plants are waste products which buildup in the tissues because there is no way to excrete them, has now been replaced. More recent investigations have led to the conclusion that many, if not all, of these substances are essential and beneficial to the plant producing them. Many are confident that, eventually, a role will be discovered for the great majority of these compounds; that organisms do not put energy and materials into processes that have no function.

Much argument, however, continues as to whether these substances are the product of a useful process or are exploited because they just happen to be there. In other words, are the chemical processes leading to formation of these substances *designed* to give the plant access to compounds which it can use in its defense or is their formation just good fortune; the fact that some are useful in defense simply accidental? Whatever the final answer to that question, what seems more certain is, that once produced by a plant, a compound useful in defense can become highly effective and can be exploited in many different ways in response to attack from predators or from disease microbes.

For millions of years, plants have had to defend themselves against grazing animals and disease microbes. It is not unreasonable to propose, therefore, that well-defended plants have always been likely to leave more survivors than those poorly defended. This trend, in turn, would ensure the survival to succeeding generations of the ability to form defense products. In response to this ability, it would be expected that predators and microbial enemies would evolve ways to overcome the defense obstacles erected by the very organisms, plants, on which they depend for survival. This 'chemical warfare' continues to the present as plants and their enemies carry on an endless evolutionary tussle to survive; a classic example of the struggle for existence.

15

Not in my backyard!

Plants tend to live in communities just as most other organisms do. Within their societies, plants compete with one another for moisture, light and soil nutrients. Therefore, many have developed a variety of ways to create *lebensraum* in their, generally, overcrowded world. The strategies adopted by plants to try and provide some breathing space around themselves by limiting competition from neighbors in their immediate vicinity are often only partially successful. Some plants smother the competition by overgrowing them, simply overwhelming and eliminating neighbors; others produce leaves very close to the ground, in a tight rosette under which other plants cannot survive, thus creating a small, tight circle around themselves within which no other plant can grow; and others produce a dense canopy of leaves thereby drastically limiting the light available to anything trying to gain a foothold in the deep shade beneath.

The main subject of this chapter, though, is those plants that use the well-known ploy to limit competition by neighbors that might best be characterized as 'attack is the best form of defense'. These are the plants which produce chemical weapons to fend off competitors.

One form of this chemical defense has come to be known as *allelopathy,* a word used to describe the ability of some plants to produce and then export into their surroundings certain chemicals which are capable of either slowing down the growth or outright killing of competing neighbors if they come too close.

The chemicals involved in allelopathy typically have simple structures; they are mostly members of either the terpene or the phenolic families (see chapter 14). The origin of these substances is uncertain; how a plant first gained the ability to inhibit its neighbors by chemical means, still a mystery. For a plant to buildup within itself substances designed to ward off attack by predators is not difficult to imagine (chapter 14); how plants might develop the trick of secreting some chemical into their surroundings to eliminate, or at least slow down, their neighbors is quite another matter.

One explanation might be that allelopathic compounds were once no more than internal defenses against predators like any other which, at some point in the past, began leaking out into the environment and, by accident rather than design, were found to affect other plants. Limiting competition in this way fortuitously, then, helped the plant producing the chemical to survive, set seed, and pass on its new-found ability to defend itself chemically to its offspring.

If we look at any community of plants we see that it is nearly always made up not just of individuals of the same species but of a number of different species, all growing together, intermingled. Look at any meadow, forest, tundra, prairie, or desert and in one small area, a variety of different and characteristic plant species will almost certainly be found growing side by side with one another.

For the individuals of any one of the species, it is not difficult to think of reasons why they might all be growing in the same place. After all, the members of a species will have very similar requirements for water, nutrition, temperature, and light, and would be expected to be able to tolerate the same environment. But why are only certain species found growing together? The different kinds of plants found together in a meadow are distinct from those found together in a prairie, for example. What determines which plant species may grow alongside others?

One possible answer to these questions has already been suggested, that is, perhaps the kinds of plants growing together have similar environmental requirements such as the amount of light they need, the temperature ranges they will tolerate, the soil conditions, and moisture levels they require. This can be seen to be true particularly in the more unusual habitats like bogs, salt marshes, alpine meadows, hot deserts, or cold tundras. Groups of plant species are characteristic of these different habitats. We can identify a bog by the kinds of plants found there. Indeed, it is likely that all plant communities are influenced by the adaptation species have for particular growth conditions.

Without doubt, plants do limit one another's growth first and foremost by competing for the supplies they need for themselves from their environment. For example, limits to soil moisture or nutrients will restrict the total number of individuals that can live within a given area. Taller members of a community may shade others from the light they need for maximum growth and slow them down. Such competition for food and water is normal in any community of individuals, whether plant, animal, or microbial.

But another aid to survival developed by plants is based on the production of chemicals which are given off to the soil and influence the welfare of other species. This type of defense has been known for a very long time among microbes. Arguably, one of the most famous examples is the discovery of Alexander Fleming who is said to have noticed the toxic effect of the fungus, *Penicillium*, on the bacteria he was cultivating in his laboratory. *Penicillium* secretes a substance into its surroundings which inhibits the growth of other microbes within its vicinity, but not itself. Fleming noticed that the bacteria growing in the same vessel as his *Penicillium* were stopped in their tracks whenever they grew too close to the fungus. His curiosity as to what the fungus might be producing that would kill bacteria so effectively led to the production of the first antibiotic, penicillin. Many chemicals with a similar property to penicillin have now been found in many different kinds of microbes. Unfortunately, by no means all of them are useful to us as antibiotics, unlike penicillin.

That higher plants, like microbes, also have a capacity to influence other organisms in their surroundings was long suspected but has only quite recently been accepted as a possibility, and not by everyone, by any means, even today.

The Swiss botanist, Augustin de Candolle, in the third decade of the nineteenth century, was one of the first to record situations in which chemical interactions between plants seemed to be occurring. For example, he noticed that thistles seemed to inhibit the growth of oats in a field; spurges, he found, interfered particularly with the growth of flax.

De Candolle not only made field observations but also carried out his own experiments. He discovered that haricot bean plants died if their roots were dipped in water in which the roots of other bean plants had been growing. He suggested that bean roots must excrete toxic chemicals into their surroundings. Similar types of tests to these were carried out by other individuals and groups through the early years of the twentieth century in which the effects on seed germination of various kinds of plants demonstrated the toxic effects of substances secreted into the soil by roots. The word ‘allelopath’ was defined in the late 1930s as a result of the work done by de Candolle and others, to mean the chemical interactions between all types of plants; some today would also include interactions between plants and the microbes that attack them, as we shall see shortly.

The idea of allelopathy in plants was recognized by gardeners and farmers long before it was understood scientifically. It was common knowledge that some plants thrived when planted together while other combinations were not very successful. The black walnut tree *(Juglans nigra)* was known since the time of Pliny, nearly 2000 years ago, to slow down or stop the growth of other species of plants nearby. The seeds of diverse species such as pine, potato, tomato, alfalfa, and cereals will not germinate if grown near a black walnut tree.

In the 1920s, it was shown that seedlings of tomato and alfalfa died when planted within about 25 m of a walnut tree. The soil zone over which the tree had influence seemed to coincide with the zone occupied by the roots of the walnut. The natural assumption was made that a root secretion of some kind was responsible for the toxic effect. This turned out not to be the case.

It was more than 30 years after these initial experiments that the correct view of the reason for the walnut toxicity was uncovered. In the late 1950s, it was discovered that the toxic effects were due to a substance leaching from the above-ground parts of the walnut tree; leaves, stems, and branches. The toxin was shown to occur in a harmless form inside the plant and to become lethal to other species only when released into the soil. The area of toxicity around the tree corresponded, therefore, not with the extent of the root system but with the area of ground underneath the leaf canopy. Leaf and twig fall, and rain caused toxic chemicals produced in the canopy to be washed into the soil and there released in lethal dose.

Other early studies of allelopathy uncovered another antisocial chemical produced by leaves of the brittlebush *(Encelia farinosa)* which grows in the hot deserts of the southwestern USA. It was noticed that, unlike its neighboring shrubby perennials which had a wide range of other plants growing in their vicinity, brittlebush was surrounded by a zone of bare ground. The toxic action of brittlebush leaves was shown to be highly specific, being lethal in tests with tomato plants but not in those with sunflowers, barley, or itself. The fallen leaves of brittlebush retain their toxicity for a year or more in the absence of rain; the toxin, which has been identified as a relative of benzene, is leached out from leaf litter into the soil by periodic rains.

One of the most careful early studies of allelopathy was carried out during World War II by the German scientist H. Bode and Belgian scientist G. L. Funke, who made observations on wormwood *(Artemisia absinthium)*. Zones in the soil about 1 m wide alongside rows of the plant within which other plants would not grow were shown to be caused by a single chemical compound, called absinthin, produced in hairs on the surfaces of wormwood leaves. The compound is washed off the leaves and drips into the soil; the toxicity of the soil is constantly renewed by rains washing fresh supplies of the toxin down from the plant.

Interestingly, chemical inhibition of growth is not always restricted to plants of different species. From time to time, examples have been found of a plant that produces a compound which is inhibitory to other members of its own species. Such an example is the rubber plant, guayule *(Parthenium argentatum)*, which, like brittlebush, also grows in the southwestern deserts of the USA. Significantly, the substance produced by roots causes self inhibition but does not seem to affect other species. The effect was first noticed by the fact that guayule trees on the edges of rubber plantations always grew better than those in the middle of the same plot. All efforts to change this pattern by extra watering or application of mineral nutrients were unsuccessful.

The guayule self inhibitor was isolated and found to be cinnamic acid, a member of the terpene family of chemical compounds. This substance is so inhibitory that it can stop the growth of guayule plants at a concentration in the soil of only a fraction of 1 percent. By contrast, tomato seedlings are only affected if treated with 100 times this concentration.

Cases like that of guayule have raised questions as to why a plant would produce a substance like cinnamic acid which is so highly toxic to its own kind and not to others. What possible advantage could there be to such a strategy?

One possible answer has to do with the way in which shrubs are distributed under the special conditions found in hot deserts. Normally, in such an environment, individuals of a given kind of shrub are widely and uniformly spaced in a seeming attempt to share scant supplies of water and nutrients. Perhaps it is an advantage to the mature guayule plant to produce a powerful inhibitor to maintain this wide spacing between individuals of its own kind thereby lessening competition for food and water.

Indeed, one of the most striking features of chaparral deserts is the zonation of herbs found around the thickets of shrubs, not just guayule but also others like sagebrush, which dominate such a landscape. Frequently, immediately surrounding each shrub are zones of bare soil some 1 to 2 m in radius. Herbs grow only outside these zones. As in the case of guayule, members of the terpene family of chemicals are generally responsible for the maintenance of these zones.

Many desert shrubs are surrounded by an invisible cloud of chemical vapors which also penetrate into the soil under leaf canopies. The compounds remain in the soil until destroyed by micro-organisms which are only active after rain has moistened their surroundings. Otherwise, the terpenes (one of which is the familiar camphor) accumulate in the soil and very effectively slow down or stop the germination of a great many kinds of seeds.

These and other examples illustrate that allelopathic effects do seem to occur between one higher plant and another. Many different chemicals are involved. What is doubtful is whether the production of any one of these chemicals is a deliberate strategy developed by a plant specifically for the purpose of preventing competition. The more likely answer is that the chance formation of a compound with toxic properties is perpetuated through succeeding generations because its production gives some advantage to the plant that can form it over the neighbor that can not.

The compounds involved in allelopathy generally occur in leaves and stems, not roots. This fact has led to much of the criticism leveled at supporters of the idea of allelopathy. Critics ask why it is that so many of the chemicals said to be involved in allelopathy are produced in the above-ground parts of a plant when it would seem more sensible and effective for them to be produced directly by the roots in the soil? So far, there is no fully effective answer to this question. The fact is that leaf and twig fall are the most important ways to deliver allelopathic inhibitors to the soil; an alternative method prevalent in arid conditions does seem to be through the formation of volatile vapors which are released into the atmosphere and then find their way into the soil.

It is, of course, competition which lies at the heart of the allelopathic interactions of plants. Competition is, after all, one of the most general and important aspects of the relationships between plants. We should view this kind of chemical warfare as just one more weapon in the wider, constant battle among organisms to establish and maintain themselves, and their kind in a community.

While it may still be unclear as to the role played by this kind of fortuitous allelopathy between one plant and another, the capacity of plants to defend themselves against attack from disease microbes through the formation of particular chemicals is much more certain. Microbial infections can be as devastating to plants as they are to animals and have often been given equally dramatic names as their animal counterparts. Scab, black leg, ring rot, blight, gangrene, and heart rot are evocative of the serious diseases they represent.

One point to remember, though, is that these exotic names reflect the diseases common among cultivated plants. Wild plants are either highly resistant to microbial attack or have fought to a draw with a disease, coexisting with it but neither overcoming it entirely nor succumbing to it completely. The wild relatives of crop plants often are much more resistant to a disease than are their cultivated cousins. In fact, breeding for resistance to disease often involves the reintroduction of genes from wild relatives into crop varieties.

Plants seem to have several levels of defense against attack by microbes. Tough waxy layers and other surface features can act as barriers to the entry of microbes into a plant but are rarely completely impenetrable. Physical barriers such as these can slow down disease attack but hardly ever stop it entirely. More effective are the impressive arsenals of chemical protectants plants are capable of producing.

The first chemical line of defense against disease lies in the presence in some plants of substances which will inhibit the very germination or early growth of microbes. These compounds are generally present at all times and are often the same terpenes and phenolics which also defend plants against invasion by herbivorous animals and insects or other enemies (see chapter 14). Many of these substances are not only unpalatable to animals but also have fungicidal and bactericidal properties. Some are so effective, such as those in older, outer onion scales and in lupine leaves, that they will prevent germination of fungal and bacterial spores more or less completely. They may be present right at the surface of the plant (leaves, for example) but may equally well be found in deeper tissues. In only relatively few cases, however, are they so effective as to prevent microbial infection entirely. In most cases this first line of defense is an imperfect one which moderates but does not prevent infection.

A second line of defense is thought to exist but is at present less well understood. A class of compounds involved in disease resistance in plants may be the, so-called, *inhibitins,* chemicals which increase rapidly the moment an infection begins. In some microbial attacks, the advance of the microbe within the plant tissues leads to the accumulation of particular chemicals just ahead of the zone of infection. Thus, potatoes infected by blight produce over 20 times more of certain phenolic compounds in a zone which lies between the area already invaded by the microbe and tissues that are still healthy.

What is not so clear is whether the production of inhibitins precedes infection or is a symptom of the disease after it has occurred. Establishing a clear link between these compounds and resistance is not so simple a matter; their presence may just be as a result of the presence of the invading microbe, not a deliberate defense against infection. The crucial question is whether the formation of the inhibitins actually does slow down the advancing microbe. The answer is not so clear in many cases.

At least in one case of a fungal disease, however, there does seem to be a reaction of an invading microbe to the inhibitins produced by the plant in response to invasion. The anthracnose fungus is a serious disease organism in cereals throughout the world. Cereals produce a number of tannins and phenolic compounds when invaded by this fungus. In turn, the spores of the fungus produce a mucilage which binds the tannins and other phenolic compounds produced by the plant in response to infection. By binding the chemicals the fungus neutralizes them and thereby overcomes the defenses of the plant. There seems no doubt that here there is a clear case of a disease organism evolving a means to overcome the chemical resistance of a host plant.

A third line of defense against diseases in some plants involves the formation of the same kinds of compounds that are known to be produced by plants to help repel attacks by enemies other than microbes. As outlined in the previous chapter, some plants store toxins which are inactive as long as the tissues they are stored in remain undisturbed. If a tissue is damaged in some way, however, either by a herbivorous animal or by microbial invasion, the toxin is released from storage.

The simplest example of this type of defense is the release of substances which then produce cyanide. A number of disease fungi cannot detoxify cyanide and are quickly killed by it. The Brassica vegetables, like cabbage, broccoli and Brussels sprouts, are examples where the wild relatives defend themselves in this way but the cultivated varieties have had the capability bred out of them for reasons of taste. Diseases produced by fungi, such as powdery mildew, are much more prevalent among cultivated than wild brassicas for this reason.

The pungent, sulfurous compounds produced by members of the genus *Allium* (that includes onions, garlic, and leeks) are one other case where the chemicals are harmless to the tissues in which they are safely stored yet are anything but benign when released after damage has occurred, as anyone can attest who has wept over a sliced onion. These, too, have antifungal and antibacterial properties.

Unfortunately, again, these barriers to infection are only partial and by no means universal. Worse still, in a few cases, the compounds in question actually *attract* rather then repel invaders. A soil-borne fungus of the genus *Sclerotium*, for instance, depends on the release of the sulfurous chemicals produced by onion roots to trigger its germination and then its infection of those same roots.

A major step forward in the search for an understanding of how plants combat disease came in the early 1940s when it was first proposed that plants contained what were termed, *phytoalexins.* At that time the idea arose that some plants, at least, seemed to form certain kinds of chemicals only at the time they were invaded by a disease organism. These chemicals, it was proposed, came into action to 'ward off' *(alexos)* disease organisms from the plant *(phytos)*.

Phytoalexins have several characteristics which set them aside from the other types of defense chemicals already described. For example, prior to infection they are not usually found in the plant at all. Only when an attack by a microbe begins within the living cells of the plant does production of phytoalexins start. They are formed rapidly, within a very few hours from the beginning of an assault by a disease microbe; their formation is restricted to the region close to where the infection begins; and they are usually effective against a wide range of fungi and bacteria.

Different families of plants produce their own kinds of phytoalexins; over 300 substances have now been identified in more than 20 different plant families and the list continues to grow rapidly. Most of the compounds are unique to a particular infection and are not found in the plant at any other time. Another striking feature of the phytoalexins is their great diversity; there is no simple relationship between chemical structure and toxicity, and no prediction can be made ahead of time as to what kind of inhibitor a plant will produce in response to an infection. Usually, too, in any given plant several substances are likely to be produced at infection, not just one.

To respond effectively to fungal or bacterial invasion, plants must quickly detect the presence of the attacking microbe and then rapidly begin producing the phytoalexins. Plants recognize the presence of an intruder by somehow sensing foreign molecules being formed by the attacker. These so-called 'elicitors' can be almost any chemical that is unique to the microbe; microbial proteins, carbohydrates, and fats, have all been found to warn the plant of the presence of an enemy. The tiniest amount of any of these compounds released by the microbe during its penetration of the plant tissues, anywhere in the plant, is sufficient to trigger a phytoalexin response by the host.

Once phytoalexin production has begun it goes ahead very quickly.

Beginning about 1 hour after an attacker is sensed by the plant there is a dramatic increase in the production of phytoalexins. Within 8 hours, there can already be more than enough of the phytoalexins present in the immediate vicinity of an attack to arrest the advance of the disease microbe.

From the point of view of resistance to disease it should be emphasized that it is not simply the ability of the plant to form phytoalexins that is important, but the capacity to produce them in quantity, quickly enough in the right place and at the right time. Speed is all important. In this and other key ways phytoalexin production seems to be the most sophisticated, effective form of defense against disease of which plants are capable.

Plants also possess other types of antimicrobial chemical defenses that seem to operate alongside phytoalexins. For example, many types of plants respond to fungal or bacterial invasion by walling off an infected area so that the spread of the disease is physically as well as chemically blocked. Other plants counterattack by producing enzymes which destroy the microbe itself. But these are generally defenses that the plant marshals more slowly so they tend to be a backup to the more rapid reaction forces represented by the phytoalexins.

But, even though there are clearly some very effective chemical barriers erected by plants in response to invasion, it is the case that disease still occurs. Virulent strains of microbes may be capable of breaching all the defenses thrown up by the plant. If and when this happens, the reaction of the plant is quite different from its previous responses. The metabolism of the host quickens as it tries to combat the advancing microbe, something like the fever produced in humans by an infection. At the same time, the microbe does not sit idly by but produces toxins of its own which are designed to poison the host, weakening its resistance. These toxins produce what we would call the typical symptoms of the particular disease whether it be wilting or yellowing leaves, stunted stems, or distorted roots.

The host plant, in turn, attempts to neutralize the toxins put out by the microbe by degrading them or diverting them into places where they can do less harm. This is the ultimate battle for control between the plant and the invading microbe and the outcome is by no means a certainty on either side. The primary defenses of the plant have been overcome by this time and have broken down; the plant is fighting for its life. It is then a matter of whether the plant can raise the pace of its metabolism sufficiently to keep ahead of the infection or whether the toxins from the disease organism will prevail and overwhelm the victim.

That plants defend themselves against invasion by disease microbes is irrefutable. Phytoalexin formation is widespread; antibacterial and antifungal compounds are also common. This kind of allelopathy clearly exists.

What is less certain is to what extent the allelopathic effect of one plant on another is a well defined strategy within the plant kingdom. This type of allelopathy does sometimes seem to operate; the walnut tree is one very conspicuous example of it. But, just as often, other plants seem to produce and give off chemicals which are as injurious to their own kind as to members of other species. Observations such as these suggest that there is no specific selection among plants for an ability to retard the growth of neighbors. Therefore, some experts believe that examples where such inhibition is found are better viewed as chance occurrences arising from the production of chemicals for some other reason, such as defense against attack by predators. Leaching of these substances into the soil surrounding a plant may or may not have an allelopathic effect. If they do, by chance, have a beneficial effect for the plant, all to the good.

The conclusion must be that plant to plant allelopathy probably does occur and may be highly significant in some cases but is not always beneficial. Its importance should not be exaggerated, at least at this point, when much evidence seems to suggest that it is a random, chancy process.

The emphatic title to this chapter certainly describes well the fierce lines of defense plants can muster against attacks from other kinds of organisms. Their defenses against others of their own kind, if they exist at all, seem ineffectual by comparison.

16

Cornu copiae: the horn of plenty

Most of us have the habit of treating ourselves when we suffer relatively minor ailments such as headaches, coughs, colds, cuts, or bruises. Our medicine cabinets are filled with more or less effective, 'over the counter' remedies to ease minor pain or discomfort. Often as not, the remedy we apply is derived, directly or indirectly, from a plant. Something like one in four of all the drugs dispensed by prescription, in our Western societies at least, are likely to have in them one or more ingredient taken from plants. Add to those the much more numerous non-prescription medications and the commercial value of these products can be measured in the tens of billions of dollars per year.

Yet, strangely, in modern Western medicine, plants are all too often seen as relics of a magical past; next to useless in the treatment of really serious illnesses. This is curious given the prevalence of plant products in common medical use world-wide. Perhaps we should remind ourselves that, in fact, plants have given to us many thousands of different medical compounds. Heart drugs, analgesics, anaesthetics, antibiotics, anti-cancer drugs, antiparasite compounds, anti-inflammatories, oral contraceptives, hormones, laxatives, diuretics, and many more are some of the categories of medicines to which plants contribute, still, today.

Of course, we should also remind ourselves that plants produce these compounds for their own use, not for our benefit. The fact that certain compounds produced by plants are of use to us as medicines is merely our good fortune. Most of these substances are formed and used by plants as defenses against being eaten by insect pests and herbivores, and to combat fungal attack, bacterial invasion, and viral infection (see chapters 14 and 15).

No one really knows how humans first came to use plant materials to cure their ailments. Primitive humans did not understand the nature of disease, viewing it as some curse of the gods or attributing it to supernatural spells. Disease had a magical and religous source, a view which lingers on today even in what we term 'sophisticated' societies. Medical quackery based on magic and superstition is by no means a thing of the past. Perhaps in earliest times, people observed the beneficial effects of certain plants on animals that ate them and decided to try them on themselves. Perhaps in trying particular plants as sources of food, humans discovered the curative properties of some of those they sampled. Whatever the origin, the interest and progression in curative plant products has today expanded until we now have a family of major industries.

The history of medicinals is the main driving force behind what today we would call *botany*. More than 4000 years ago, several hundred drug plants were already known to the Assyrians. Slightly more recent Egyptian records show that the preparation and regular dispensing of medicinals was common practice in that early civilization as well. The ancient Greeks gave close attention to medicinal plants and by a little over 2000 years ago the famous Dioscorides, physician to the Roman army, had compiled a detailed account of several thousand plants in his *De Materia Medica,* a book which remained the authoritative reference on medicinal plants for the next 15 centuries, in the Western world at least. Throughout all that time, botany was mainly just the compilation of more and more examples of plants useful as medicinals.

The Dark Ages were, in nearly every way including medicine, a long period of ignorance in Europe. Indeed, until relatively recently people still had not much more knowledge of how disease worked than had their prehistoric ancestors. Such entrenched and ancient beliefs as the 'Doctrine of Signatures' was widely held in the Middle Ages in Europe; that like cured like. The Doctrine assumed that all plants were created for the use of humans and were endowed with certain forms and shapes that marked them for a specific use in treating similarly shaped organs in the human body. The walnut was a brain tonic; bloodroot, a blood tonic; heart-shaped plants could cure heart afflictions, and so on.

The appearance of the herbalists and apothecaries, in the centuries from about 500 to 300 years ago, heralded the advent of modern botany and, hand in hand with it, modern medicine. The 'herbals' from which these forerunners of the medical and pharmaceutical professions of today studied and practiced were magnificent botanical volumes full of artistry, beautiful illustrations, superstition, magic and, most importantly, the beginnings of observational science – the systematic testing of the effects of medicinals on particular human diseases and conditions. Modern medicine slowly emerged from this ancient, murky, muddled background.

From about 200 years ago to the present we have made startling progress in our understanding of the curative properties of plant chemicals and how they act.

Unfortunately, the pendulum has swung rather too far against plants which are often viewed now in Western medicine as being, as I said earlier, relics of this magical, quasi-religious past. Most of the old remedies of the herbalists and apothecaries have been discarded now in modern Western medical practice leaving behind relatively few still in use. The 120 or so drugs currently in widespread use that are based on plant products come from less than 100 different kinds of plants. This is mainly because, of the quarter of a million species of flowering plants known world-wide, only about 2 percent have ever been tested to see if they contain compounds of medicinal value. Even among these, the testing has been very limited. For example, it is not uncommon for a plant to be assessed just for its value in combating a single disease or condition. A pharmaceutical company interested in trying to find a cure for acquired immune deficiency syndrome (AIDS), cancer or malaria may test a series of plants only against the disease of interest to them. Not often enough are extracts from plants put through a battery of tests covering a wide range of medical conditions. Perhaps it is not surprising, therefore, that very few of the plants tested in this way have ever been acknowledged in Western medicine to have any real therapeutic value. Add to this the fact that it can take upwards of 100 million dollars to learn how to extract and manufacture a new drug and 10 years to put it through the tests required before it can be released on the market. Under the stringent conditions demanded by modern society, progress in finding and marketing new drugs is bound to be painfully slow.

Western medicinal methods, however, are used to treat only a tiny proportion of all the earth's peoples. Some 80 percent of the world's population relies entirely on local medicines; these are made almost exclusively from plants. Looking at the matter from this perspective, it is thought that as much as 30 percent of plants have been used as medicines by various peoples. As the cost of Western medicines continues to rise beyond the economic reach of poorer countries, this percentage may very well rise.

Even within Western society, the increasing popularity of such fields as homeopathy and other forms of, so-called, 'non-traditional' medical treatments is an indication that drug therapy, as we have developed it in the West, is not so widely regarded as it was a few years ago. Rightly or wrongly, there is a feeling that our inability to accept the observations and practices of traditional healers has prejudiced our recognition of the fact that there may be many more useful plant medicines than we have led ourselves to believe. We should, perhaps, remember that even the way some of our Western drugs work is only poorly understood. We still have little idea about why aspirin seems to be effective against so many, diverse medical conditions, for example. How much more so are we ignorant of the more complex mixtures of curative concoctions from plants traditionally used by the shamans and witch doctors? A little humility may lead us to some valuable insights into the ways in which we might again use the traditionalist approach to medicine to our advantage in the future.

Modern searches for plants containing useful medicinals are based on tests that only allow detection of compounds that affect particular medical conditions. Thus, a test for AIDS may be developed, for example, and plant extracts tested for their ability to combat this disease. Tests for other, single diseases are used to screen extracts in the same way. A pharmaceutical company may spend many years and millions of dollars in such searches and find nothing that works in the particular tests they carry out. Clearly, by no means all of the quarter of a million flowering plants, plus the thousands of conifers, ferns, mosses, and so on, could possibly be tested in this manner. Just collecting the samples is a major undertaking since it may involve sending armies of collectors out across the world to the remotest locations (tropical rainforests, for instance) at huge cost. Nonetheless, two strategies have been used in these attempts; the 'random' and the 'targeted'.

In the random type of search, a range of plants is simply collected from likely locations and extracts from them tested; no other criteria are applied in the screening process. Such sweeps around the globe were common in the 1960s but were carried out often by ill-trained people who knew little or nothing about medicine. Consequently, the success rate was dismally low and led to much cynicism about the value of plants as sources of medicines.

Targeted surveys take several different forms. Sometimes it is useful to look for relatives of plant types known already to produce a useful compound. In this way, even more effective variants of already effective chemicals may turn up or, alternatively, plant types which have in them greater amounts of a sought-after compound.

Sometimes, more precise ecological searches yield results. For example, if plants are sampled from an area where slugs and snails are major predators, compounds which combat attack by like organisms may be found. These compounds may turn out to have value in combating diseases like schistosomiasis, a disease of central Africa spread by snails.

Finally, of course, are the searches based on the traditional medical practices of indigenous communities. Here, the difficulties are great but the rewards can also be significant. It is not easy to gain the trust and confidence of people so that they will reveal their most precious traditions. It is nearly impossible simply to walk into a society in some remote location and at once expect to be showered with information by shamans whose power often depends on the secrets they have learned from previous generations of traditional medicine men. Years may have to be spent gaining the confidence and cooperation of such communities. Even then, success is not assured. Despite these difficulties, the history of drug discovery and development seems to suggest that this type of targeted search is the most rewarding, in the long run; at least far better than random screens.

Plant chemicals are used in three main ways in Western medicine nowadays: some are used directly in medicines and medications; others are analyzed so that their exact chemical structures are known and then ways are found to manufacture them artificially rather than just relying on the natural source, which may be rare, anyway; and finally, they may not be used directly in treatments at all but as tools to help us develop new drugs. For example, a plant may be found which has in it a compound which has only a mild effect on some disease. But, by extracting and working out the structure of this compound, and manufacturing it artificially, we can then change its structure in several different ways to see if we can improve its effectiveness or gain new insights as to how to design a more effective drug of a similar type.

Very few types of medicinal plants have remained in favor for even so much as 100 years. Some – opium, cocaine, marijuana, and quinine are four examples – have survived the centuries but most are much more recent.

One of the most renowned botanical drugs is opium from a species of poppy native to Asia Minor. Legends about opium are recorded in Egyptian hieroglyphs and Homer in his Odyssey speaks of nepenthe, a substance that would 'lull pain and bring forgetfulness of sorrow'. At the same time, others warned of the addiction to *opos,* the milky sap of the poppy capsule. Many down the centuries have written with fascination of this great analgesic.

The main attributes of opium are mostly due to the morphine it contains. Yet, morphine is just one of some 25 medically interesting compounds found in the milky latex exuded from the immature seed capsules of the poppy when it is cut open. Morphine takes its name from the Greek god, Morpheus, the bringer of dreams. The Greeks regarded sleep as the great healer; the opium poppy, therefore, was celebrated in their art, literature, and religion. Long before the ancient Greeks, as far back as 8000 years ago, there is evidence of small opium balls being eaten or mixed with wine to induce sleep and relieve pain. In nineteenth century Britain and elsewhere, laudanum (now called tincture of opium) was regularly prescribed by physicians, causing widespread addiction to the drug.

In the latter half of the nineteenth century, chemists began to tinker with the morphine molecule to try and make a compound that would be more effective than the natural product as well as less addictive. By the end of the century, German chemists had, indeed, found a more effective painkiller made by modifying morphine. Its name? Heroin!

Codeine, another of the compounds found in the latex of the poppy (but only one-fifth as strong as morphine) is still widely used as a household painkiller, especially in cough medicines. Incidentally, heroin once was regarded as a much more effective cough suppressant than codeine and was openly sold in cough medicines, in North America at least, until near the end of the second decade of the twentieth century. It is suggested that there may have been as many heroin addicts in New York City in the early 1900s as there are today. Most of them, however, were children treated for coughs, often the result of tuberculosis in those days.

Hospital patients everywhere have benefited greatly from the leaves of the coca bush *(Erythroxylum coca)* which have provided some of the most valuable and widely used anaesthetic compounds in use today. The leaves contain some 14 different ingredients, including cocaine which itself is used as an anaesthetic but which also has been used as a blueprint for the development of a number of artificial painkillers such as procaine and novocaine.

Besides giving us cocaine, the coca leaf is famous as the stimulant used by Andean porters and laborers who, reputedly, can work with superhuman endurance for days with little or no food if they chew coca leaves. When chewed with a little lime or plant ash, active compounds are released from the leaf which stimulate the nervous system enhancing muscle potential, increasing stamina, depressing hunger, and relieving pain. Distances in the Andes can be reckoned in 'cocadas', the distance that can be traveled on one coca leaf chew.

Interestingly, since the South American Indians do not take in cocaine alone, but only as one ingredient of many in the coca leaf, they do not suffer the addictive and mind-altering side effects we commonly associate with the pure compound. Why this is so remains a mystery. Pure cocaine we know is dangerously addictive whilst cocaine in a crude mixture taken directly from the coca leaf appears not to be so to nearly the same extent; just one more example that there is sometimes more to natural plant medicines than we give credit for.

One more ancient drug that today we judge to be too dangerous for general use, described by the Chinese nearly 5000 years ago, is marijuana (also known as hashish, ganja, or kif) made from the resinous oil of the cannabis plant. Experience seems to show that marijuana is not particularly addictive and less habituating than alcohol or nicotine. Hashish, the unadulterated resin from female cannabis flowers, is most famously associated in folklore with a curious Mohammedan sect, the 'hashishins' (whence 'assassins'), whose avowed religious purpose was to murder 'enemies of the faith' in order to enter paradise. Hashish figured prominently in arousing their religious fervor, and the sect numbered in the tens of thousands in Syria and Persia until they were violently and forcefully suppressed by the Tartars in the thirteenth century. Arguments for and against the legalization of marijuana continue to the present day, so far without final resolution.

Of course, not all drugs derived from plants are as currently socially unacceptable as the three discussed so far. One of the most remarkable of those in the 'approved' category is quinine. A small, attractive evergreen tree with glossy leaves and fragrant pink or yellow flowers provides the world with one of its most precious medicines. This drug, found in the bark of several species of the cinchona tree, which are native to the humid, forests high in the Andes of South America, is still the foremost cure for malaria, a disease which affects over 100 million people a year.

This antimalarial drug was used by the Andean Indians long before Europeans discovered the, so-called, New World. The conquistadores and the Jesuits brought it to Europe from South America at the beginning of the seventeenth century. However, the entrenched and powerful medical professionals in Spain and elsewhere ridiculed and rejected the cure since they were making significant financial gains by treating patients suffering from malaria with their own patent remedies. Then, Robert Talbor, a British apothecary who lived and practiced at the end of the seventeenth century, successfully cured King Charles II of the 'ague', as malaria was then known, with bark from a Peruvian cinchona tree. From then on this miracle cure was elevated to the center of attention in Europe.

By the second decade of the nineteenth century, quinine itself had been isolated by French doctors and named from the Amerindian word for the cinchona tree, quinaquina (which means 'bark of barks'). The demand for the drug then rose so dramatically that by the mid-nineteenth century, wild cinchona trees were being exploited to extinction. Even their roots were being dug up and stripped of bark.

Fortunately, this vital plant component in the world of medicines was rescued by a British seed collector, Charles Ledger, who sent some cinchona seeds to Europe. The British government declined to buy the seeds; however, they were purchased by Dutch authorities for about six pounds sterling. From 450 g of these seeds, the Dutch grew 12 000 cinchona trees in their colony, Java, and for the next 100 years controlled nine-tenths of the world's supply of this vital medicine. The original purchase of seed by the Dutch has been described as the one of the best investments in the entire history of commerce!

The world monopoly in quinine supply by the Dutch was broken by the World War II invasion of Java by the Japanese. Seeds were flown out of the colony by the Dutch at the last minute but this did not help the troops already serving in the Pacific theater of conflict. The USA, instead, sent a team to Columbia where nearly six million kilograms of dried bark were collected for use throughout the war.

Synthetic quinine was finally manufactured in the USA towards the end of World War II and from then on the demand for the natural product declined. Interestingly, however, in the 1960s, certain strains of the malaria parasite were found which were resistant to this synthetic form of the drug. There are today areas of the world where the synthetic drug is not used at all because of this resistance. But none of the strains of malaria parasite is yet resistant to natural quinine. The natural product (often called 'totaquine') has in it four antimalarials, quinine among them, not just one as in the case of the synthetic drug. This seems to be another outstanding example of a natural medicine being superior to an artificial copy, similar to the case of the coca leaf versus pure cocaine; both are examples where a mixture of natural products out-performs the purer, but synthetic counterpart manufactured in the Western medical fashion.

Arguably, the most widely used drug world-wide is aspirin. Though the compound today is made synthetically, it was substances extracted first from the white willow *(Salix alba)* and later from the herb, meadowsweet

(*Filipendula ulmaria* as it is now known) which gave to chemists the blueprint for making this universal curative.

The willow has a long-held reputation as a painkiller, certainly since the days when the Doctrine of Signatures was in vogue. The observation that willow leaves, as they moved in the wind, seemed to resemble the trembling of a fever patient led to the ancient use of infusions of white willow bark as a treatment for fevers. Yet further back in history, as early as 2000 years ago, at the time when Dioscorides was compiling his *De Materia Medica,* the willow was also being used to treat gout, rheumatism, toothache and earache, as well as headaches.

Eventually, it was found that the active ingredient in meadowsweet, salicylic acid, was more effective than the related chemical compound first extracted from willow. By the end of the nineteenth century, German chemists had discovered that salicylic and acetic acids could be combined to produce acetyl salicylic acid (the now familiar ASA) which proved more potent than anything that had gone before in relieving all kinds of pain. The trade name for aspirin, thus, was derived from the nineteenth century Latin name for meadowsweet, *Spiraea ulmaria,* not from willow. The medical conditions that appear to be helped by aspirin seem endless; the list continues to grow every year. Yet, even after decades of intensive investigation, we still have little idea why the compound is so effective in so many ways.

A somewhat more recent major discovery of a plant compound of medical interest than those discussed so far occurred in the 1940s. One of our best known forms of birth control, 'the pill', is derived from diosgenin, a compound found in high quantity in yams, that belong to a family of climbing forest vines. When American chemists showed how diosgenin could be used to produce male and female hormones, the sexual revolution which is still going on today was launched.

The same group of plant chemicals from which diosgenin comes also contains compounds that are the starting material for other important families of drugs used in the world today. Alongside oral contraceptives we can place cortisone and hydrocortisone as well as sex hormones and the anabolic steroids. The current widespread abuse among athletes of some of these products should not detract from the fact that many of them have widely beneficial, legitimate medical value.

We tend to think of garlic more as a condiment than as a medicine; we should consider it as both. As a natural antibiotic, garlic has been used in many areas of the world to treat, especially, nose, ear, chest, and throat infections. For the early treatment of colds and flu, cloves of garlic kept in the mouth is a long-standing folk remedy. Sufferers from colds, flu, sinus congestion, whooping cough and bronchitis as well as high blood pressure, acne, asthma, and even diphtheria have all benefited from the use of garlic.

As early as 5000 years ago, Mediterranean and Far Eastern peoples were already using garlic. The Greek physicians, Hippocrates and Galen, prescribed garlic for various infectious diseases; Dioscorides used it to treat worms in soldiers of the Roman army.

The active ingredient of garlic is called allicin, a compound which has a distinctive smell and which acts by literally suffocating infectious bacteria. Nearly all the members of the lily family related to garlic, such as onion, leek, and chive, also have some form of medical value. Onions in particular, other than garlic, have been used for more than 5000 years and were for a long time highly prized for their medicinal properties. They are said to stimulate insulin production, lower cholesterol levels in the blood, promote healing of wounds, and suppress allergies. Today, we take them for granted and use them only in cooking, not especially as curatives.

Today, a major focus for the development of new plant medicines is the rainforests of the world. There, plants are in close competition with one another for survival; for space, light, and nutrients. Rainforest plants also must protect themselves from attack or injury from a host of insects and other voracious predators. In order to survive in this cutthroat world, many types of plants have developed poisonous chemicals which are designed to devastate enemies (see chapter 14).

We have only just begun to analyze and medically evaluate compounds from our rainforests systematically. Not much more than about 1 percent of the plants in the Amazon rainforests, for example, have been sampled so far. New chemicals are being discovered in their dozens but few so far have been shown to have promising medicinal properties. At the same time, of course, the rainforests of the Amazon and elsewhere are being destroyed at a rate which is greater than the pace of evaluation of the plants found there. Potential remedies for illness are likely being eliminated; we shall never know for sure.

One of the most important, relatively recent rainforest additions to the list of medicinal plants is the Madagascar periwinkle *(Catharanthus roseus)*, discovered in the early 1960s. More than 100 chemicals of some interest have been isolated from this plant so far. The most important, medically, are vinblastine and vincristine, both used in treatment of certain cancers. Vinblastine offers a cure rate of better than 80 percent among those struck down by Hodgkin's disease while vincristine produces a remission rate of about 90 percent among children suffering from acute and lymphatic leukemia.

One current, intensive search is for plant products which could aid sufferers from arthritis. Millions of people around the world suffer from one or more of as many as 200 different types of this debilitating illness. More and more sufferers are seeking relief by turning to homeopathy, with some success. Reportedly, an oil extracted from the evening primrose (*Oenothera* species) shows promise for improving the condition of significant numbers of arthritis patients. This oil has also been used to help hyperactive children as well as sufferers from migraine, asthma, eczema, hypertension, premenstrual syndrome, Parkinson's disease, and multiple sclerosis. Evening primrose oil may be one more significant plant product in the near future in Western medicine.

Only a few of the more familiar medicines isolated from plants over the centuries have been touched on here. The list is a very long one and could include other well-known substances like eucalyptus, camphor, wych hazel, liquorice, belladonna, senna, and ipecac, to mention only a few more of great value world-wide.

It would seem that the contempt of some for plant cures in modern Western medicine is misplaced. Not only are some of the oldest of our ailment remedies still among the most popular of our curatives but we are also rediscovering in the West something fortunately not forgotten in other areas of the world; that plants can help to keep us healthy. They may, indeed, offer the only hope for the future in cases where the useful chemicals they contain are too complex or difficult to manufacture artificially. Regarding green plants as *cornu copiae* of medicinals is more of a truth than we have recently been willing to admit. Mother Nature is, arguably, still the best chemist around!

17

Getting dead

In the preceding 16 chapters, I have attempted to present a broad account of how, what are called, the higher green plants go about their daily lives. The first six chapters deal with how plants gain the water, nutrients, and energy they need, and how they move essential materials around from roots to leaves and back again in their, sometimes, enormously large, sprawling systems.

This group of chapters is followed by another five dealing with growth and development. Here, how plants respond to cues from their environments (changes in light, temperature, and gravity) are set beside other discussions of how they sense day and night, how they respond to the changing seasons, and how they deal with or, in some cases, avoid daily and seasonal stresses placed upon them because of their sedentary lifestyle.

In a final group of five chapters, discussion centers on how plants use their unusual capacity for producing an enormous range of exotic chemicals to attract pollinators and seed dispersers by adding color, fragrance and flavor to their flowers and fruits; defend themselves against attack from predators and diseases and encroachment by other, competing plants in their immediate vicinity; and how we humans have exploited the marvels of natural chemistry in pharmaceutical, cosmetic, and a host of other industries.

So, what has been left unsaid? Individual readers may be able to add their own topics to those covered here; I could probably add one or two myself. But one stands out above all other possibilities.

The entire book so far has dealt with the life of the plant; how its systems work to promote and sustain life. What about death, then? How do plants 'get dead'? I do not mean by this what has become one of the most frequent queries from the public to commentators on radio and television 'garden hot line' programs – 'Why is my (house, garden, favorite) plant dying?'. The word 'neglect' is probably the commonest (but unspoken) answer to that! Rather, why and how do plants die in the wild? Do all plants die, inevitably? What do we know about the processes leading to death in plants? This is the topic of this final chapter – to discuss something of what we understand about how plants get dead.

If we think of death and dying at all, and most of us do so as infrequently as possible, we tend to imagine it as a process which begins at birth and progresses to an inevitable end, sometime. The phrase 'life span' succinctly summarizes how we view the matter. To us, there is a beginning and an end to life. Mind you, a life span is actually the maximum length of time an organism *could* live if all the conditions of life were at their most favorable. The human life span, for example, is thought to be about 120 years. To be realistic, most of us do not expect to be around for anywhere near that long. Longevity, for most of us, more pertinently describes our hopes and expectations in life.

At his 100th birthday party, the American jazz musician, Eubie Blake, reportedly said that if he'd known he was going to live this long, he'd have taken better care of himself. In making this wry joke, Mr Blake was, probably unwittingly, referring to another way of looking at life and death – in terms of longevity. Here, the question is not 'How long *could* I live?', but, rather, 'How long can any individual *expect* to live?' This is quite a different matter from life span. How long we can expect to live depends on prevailing environmental and cultural conditions. In some parts of the world today, for example, human life expectancy may be still only 30 or 40 years, about the same as it was some 2500 years ago at the height of ancient Greek culture, whereas in others we know it to be close to 80 years. Disease, predation, accident, flawed lifestyles, and polluted environments are just some of the hazards to life faced by ourselves and all other living things. Longevity is sometimes starkly different from a theoretical life span; how well we look after ourselves during our lives can have a profound influence on the former but not on the latter.

Regardless of the issue of the longevity of individuals, all types of organisms face one absolutely basic event in life, namely, reproduction. Death must be delayed long enough to allow, as a minimum, at least one opportunity to breed. Thus, the survival of at least some individuals to reproductive age is crucial to the passing on of genetic traits to descendants. If a way were not found to allow this transfer of genes to the next generation, then any particular kind of living thing would be truly dead, and within one generation, too!

So, regardless of the extent of longevity or what is meant by life span in the case of any particular species, what is important is the different strategies living things have evolved to achieve the fundamental aim of ensuring the advent of the next generation. These we call *life history strategies;* in plants they can vary enormously.

Some kinds of organisms appear to be immortal. As long as they have a proper food supply and a favorable environment, some living things do not have to die, nor do they do so. Bacteria, for instance, simply divide in two; both 'offspring' continue as did the 'parent' until they, too, divide in the same way, *ad nauseam.* Some types of animals like sponges have the ability to fragment their bodies and release individual cells each of which is capable of producing a new sponge. Some plants, such as a number of the grasses, like the buffalo grass *(Buchloe dactyloides)* of North America, or bracken, as another example, which spread by underground stems, seem to have the ability to go on growing indefinitely. Some ancient colonies of these species are known which may have propagated themselves, in an unbroken line, since shortly after the end of the last ice age; in other words, they may be as much as 15 000 years old and still going strong! Some soil fungi, which spread underground as fine filaments, have perpetuated themselves in this same way, we suspect for many hundreds of years.

Other examples of a similar kind are some of the antique garden plants which have been continuously transplanted over centuries using small pieces of the plant (grafts, runners, 'slips', and rhizomes). Some apple varieties and grape vines in Europe and cultivated olives in the Middle East, for example, have been propagated in this way for hundreds or, in a few cases, even thousands of years.

We call organisms perpetuated in this way directly from the parent, *clones.* All clones are biologically identical to the original parent, whether they be plants, animals, fungi, or bacteria.

The terms longevity and life span in the case of clones lose much of their meaning, therefore. Death is not an inevitable consequence of life; certainly not of *all* life. Thus, not only can the longevity of individuals in a population vary enormously depending on environmental and cultural conditions, life span from one kind of organism to another also can vary, all the way to infinity, as far as we can tell! Given this enormous flexibility, it is not surprising, then, that we find a dizzying array of different life history strategies among living things.

Some of the oldest known living individuals are plants. The best verified cases we have are the bristlecone pines *(Pinus longaeva)* found in California, some examples of which are now known to be within a few decades of 5000 years old. At the other extreme are some of the desert plants, the aptly named 'ephemerals', which germinate from seed only after a significant rain, then grow, flower, set seed again and die in a matter of a week or two. Clones aside, plants seem to have the widest range of life histories of any of the kingdoms of living things.

They may vary very widely in their strategies, nonetheless, the flowering plants can be placed in one of two quite distinct groups based on their life histories. On the one hand are those capable of reproducing many times during their lives. Most plants of this type, once they reach maturity, alternate between periods where they put all their energy into producing more roots, shoots, and leaves, and other periods when they produce flowers and, then, seed. These *polycarpic* (literally, bearing fruit many times) plants are typified by such examples as trees, shrubs, perennial grasses and those, like tulips, irises, and gladioli, that have bulbs or other kinds of storage organs.

The remaining flowering plants are the *monocarpics* which bloom and set seed only once in a lifetime, then die. All plants that live just one (annuals) or two (biennials) years are of this type. Biennials, typically, form only leafy growth in their first year before flowering and setting seed in the second. A few perennials are also monocarpic, the most spectacular examples being species such as the giant *Puyas*, century plant, talipot palms *(Corypha umbraculifera)*, or one of the bamboos, which can grow for upwards of 100 years, or even more, before flowering just once and then dying. The most conspicuous examples of monocarpic life strategies are, however, our agricultural crops where whole fields of plants, like our cereal grains, germinate, mature, flower, set seed, and then die, all at much the same time. Plants of this kind provide conspicuous examples among living things of what is called *senescence,* a built-in program leading inevitably to the death of an entire organism or some part of it.

Mass, programmed senescence and death of the kind typified by agricultural crops is not confined to plants but is also common in the animal kingdom. The Pacific sockeye salmon, for example, lives for 7 years, spawns once and then dies; mayflies, among many kinds of insects, swarm, mate and then die *en masse;* and the drones in a bee colony attempt to mate with a queen after which they die.

But, in most of the more complex animals, deterioration of the whole organism is gradual. It is not easy in cases like these to pinpoint what exactly contributes to eventual death; a program of senescence is not obvious. In ourselves, for example, bodily functions certainly decline with advancing age. These inexorable changes affect almost every one of our systems from the wrinkling of skin, to declining renal function, cardiac malfunctioning, and decreased physical and mental activity. There are tight connections among the organs in complex animals like humans so that even a minor malfunction in one of them can contribute significantly to the decline of the whole organism. But the onset and progress of the multitude of symptoms that we call 'aging' are very uneven among any group of individual animals so that it becomes difficult to predict with any certainty which ones *will* contribute to eventual death. There is no inexorable, sequential, predictable program of decline preceding death.

Of course, plants age too. In fact, perennial and polycarpic trees and shrubs show the same kind of aging and death pattern as animals in the wild, where few individuals live long enough to die of natural causes. In these cases, there is generally a steady culling of individuals, through infant mortality, disease, natural disasters, or the activities of predators and browsers. Only a few individuals reach old age. Death in these survivors may follow, from the accumulation of toxins, for example, which may have builtup over many years. But aging of this kind should not be confused with senescence; they are two different things.

For example, in the case of some ancient trees, a twig can be removed, rooted, and a new sapling raised which has the same degree of vigor as a fresh seedling. The aging of the parent tree in these cases is likely linked to regressive changes, such as the decay of the trunk after decades of wear and tear or the inability of the canopy of leaves any longer to provide food for the ever increasing volume of non-green tissues of the plant, not to any decline in vigor, the programmed senescence of parent shoots. Often, growth of individual parts of a tree or shrub is only loosely tied to what is going on in the rest of the plant. The supply of water and nutrients may be important but even restriction of these essentials does not necessarily cause inevitable, irreversible damage to the living tissues of plants. The tree may be aging but its growing tissues are not necessarily senescing. As for complex animals, a program of senescence leading to death of the whole plant is not easy to discern in cases such as these.

Much of the information gained about senescence of the whole plant has been discovered through the study of annuals. Annuals have two great advantages over other plants. In the first place, they have the potential to multiply and spread very quickly since they tend to breed and then produce large amounts of seed in a very short time. Secondly, their life strategy allows them to reach the next seed generation in only a few weeks or months. This, in turn, allows them to survive adverse climatic conditions in a dormant or quiescent state in the soil.

Weighed against these aids to survival among annuals are one or two quite severe constraints. For example, it is difficult for annual plants to grow sufficiently big in one season to compete with larger, bulkier perennials. In other words, annuals tend to be 'shaded out' wherever they grow if they are forced to compete with longer-lived plants. This we know from our own gardens where we try not to plant our annual flowers or vegetables in locations which are likely to become more shaded as the season progresses.

In the wild it is also difficult for annuals even to establish themselves unless there is relatively bare ground in which to do it. Dense vegetation will generally choke out annuals which, therefore, thrive best in conditions where competition from other plants is much reduced. These more favorable growing conditions can be created by the removal of other plants through cultivation, by the activity of burrowing animals, or because of the felling or burning of trees in a forest to produce clearings. Annuals are often the first colonizers to reappear after a forest fire when competition from large perennials has been drastically reduced. Their dominance does not last long, however, once longer-lived plants begin their comeback. Desert ephemerals have a natural advantage of this kind in their sparsely populated environment. The competition for open space is not very severe in places like deserts where most of the ground is bare of vegetation most of the time.

Because of constraints such as these on annuals, it is vital to their continued well-being to have built into their life strategies a healthy degree of 'bet-hedging', especially in connection with their seed germination. It would be suicide, for instance, for an annual plant to have all its seed crop germinate at the same time or in an environment where conditions were likely to deteriorate so badly that all the seedlings would die. Dormancy strategies have arisen in annuals, therefore, which tend to spread out germination of any single seed crop over several seasons (see chapter 10).

If annuals depend on bet-hedging for survival then plants which spend one or more years in a leafy condition before flowering once and then dying can be called 'big bang' strategists. At one end of this spectrum are the strict biennials, like carrots, sugarbeet, or onions among garden vegetables, which flower and die in their second year of life, if they are allowed to complete their natural life cycles. At the other extreme are long-lived species like the silversword (*Argyroxiphium* species) from Hawaii, which grows for about 7 years by which time its rosette of leaves is about 60 cm in diameter; then it flowers and dies. A *Puya* species from the Bolivian Andes flowers only once after about 100 years and promptly dies; some of the *Agave* species and one type of bamboo do likewise. All plants in this category make one supreme effort to flower and set seed at some point in their long or short lives, then they die.

The perennials which produce leafy growth and flowers every year for either a few or many seasons are a particularly diverse group with a wide variety of life strategies of their own. They are typified by the fact that they occupy a certain space year after year but die back to the ground at the end of each growing season. The advantage of such a life strategy is that the shoots are much less vulnerable to problems of drought, cold (or heat), or predators, for example, during an unfavorable season. They generally have storage organs below ground which see them through to the next period favorable for growth. The main disadvantage, compared to the woody shrubs and trees, is that at the end of each season they surrender their position in the canopy of plants competing for light; at the beginning of each new season they must re-establish themselves quickly by producing again all their above-ground parts.

Weighed against this apparent disadvantage, however, is the fact that these leafy perennials need not divert their energy and other resources into the production and maintenance of permanent woody, supporting structures like the shrubs and trees do. Many of them also have the ability to 'move around' through the formation of rhizomes, stolons, or shoots that arise directly from roots. In this way, plants like the grasses and sedges, for example, can spread over large areas to form intricate clonal stands; alternatively, they can 'jump over' barriers like uninhabitable patches of ground by simply growing across them to more favorable territory beyond. Any gardener who has tried to rid a plot of land of a plant that grows in this way, by producing rhizomes, stolons, or through spreading roots, knows the near impossibility of eradicating such 'guerrillas' entirely.

The larger woody shrubs and, especially, the trees tend to delay investment of energy and resources in reproduction until sufficient growth has taken place to ensure their survival for a long period of time. Once they have established their 'place in the sun' plants of this kind can then devote some resources periodically to flower and seed production; those individuals that do not establish themselves in the high canopy will usually die without leaving offspring.

A general conclusion that might be drawn from a study of the wide variety of life strategies among plants is that the shortest-lived kinds (but also including a few long-lived monocarpics) tend to have their lives limited by programmed senescence. The perennials, on the other hand, tend to have much less precise limits to their lives and tend to die through a gradual attrition associated with aging.

Yet it is also true to say that no plant is completely without some form of senescence. Senescence in plants shows a wide range of patterns. The total death of the entire plant at the end of reproduction, as in all the monocarpics, is the most extreme case; somewhat less drastic is the death of all above-ground parts (top senescence) at the end of a growing season, a form characteristic of plants with storage organs like bulbs; and less drastic again is the senescence of an entire leaf array at the end of a growing season, leaving the stems and roots bare for a period of time but alive (deciduous senescence). A modified form of deciduous senescence is the periodic fall of leaves and small branches characteristic of evergreens. The needles of many evergreens last about 4 years, on average, although in the case of the extremely long-lived bristlecone pine, 30 years is more common. Still less drastic is the progressive senescence of leaves along a stem, the dieback of the oldest leaves in the normal development of annual plants (progressive senescence).

All these patterns, however, point to the same conclusion; that senescence must be under some kind of control within each plant. Whether the program of control is the same or different in each case is also a question. What, in fact, do we know about the processes which lead to senescence and death in plants? The short answer is, very little, despite many decades of searching for answers.

For centuries, gardeners and horticulturalists have known that it is possible to extend the life of an annual plant by constantly removing flowers and fruits. If the wish is to extend the period of blooming in garden plants, fading flowers should be removed before they die and are replaced by fruit. By doing this, it is possible to extend the life of monocarpic plants, most of which normally live only a few months, to several years. In soybeans, for example, which are annuals, it has been possible to induce plants to grow to a height of about 8 m over a growing period of 15 months simply by removing flower buds as they develop. Soybeans can also be prevented from flowering altogether by maintaining the plants under artificial light in a greenhouse in long day and short night conditions. Soybean is a plant which will flower only if exposed to short days and long nights (see chapter 9). Otherwise, if given long periods of light and short lengths of darkness on a daily cycle, the plant will never flower. Preventing flowering, and then seed production, in this way causes soybeans to continue leafy growth for years without dying. Some other annuals can also be kept alive for decades as long as their flowers are not allowed to mature fully.

Observations like these point to the fact that, at least in the case of plants which die after flowering just once, the control of the events leading to death depends not so much on what is going on around the plant, in its environment, but on controls within. These controls are also linked, somehow, to flowering and seed production. Nipping flowering in the bud or preventing it from starting altogether seems to be the key to delaying death, sometimes for years.

The question then becomes what within the plant acts to trigger the processes leading to senescence and death? Is it that flowers produce some kind of signal once their role in setting seed is fulfilled?

It is important to appreciate that senescence is controlled by a genetic program found in each living cell of every individual. Using leaves as an example, senescence begins there with the loss of green color followed by the export to the stem of much of what the leaf contains. Only a brown, yellow, or red husk remains behind to be shed, eventually, in 'leaf fall'. Valuable components, like carbohydrates, proteins, lipids, and other chemical materials contained in the leaf, are first mobilized and then moved out into stems and roots to be used again later either in flower and fruit formation or elsewhere in the plant. All of this processing within the dying leaf takes place in each of its living cells. The same can be said of any form senescence takes; the process is inherent within each living cell.

Not surprisingly, therefore, given the fact that so much of the contents are mobilized and moved out, one attempt to explain what triggers leaf senescence is that flower and fruit development causes competition between these reproductive organs and other parts of the plant, such as the leaves, for the nutrients essential to growth.

One striking feature of monocarpic plants, certainly, is the sharp shift in the movement of nutrient resources (minerals and carbohydrates, for instance) away from leaves, shoots, and roots towards newly formed flowers and fruits. The growth of roots and stems, and the production of new leaves often decreases in plants of this kind and stops altogether soon after flowering begins. One explanation is that this shift in nutrients towards flowers and fruits acts as the signal for the onset of senescence.

Diversion of food and other nutrients away from leaves towards flowers and fruits cannot be the complete explanation of senescence, however. Most plants do not form large enough numbers of flowers to cause such massive diversions of nutrients away from leaves that would result in the latter starving to death. In any case, in some plants, even if flower buds are continually removed, leaf senescence still occurs. In other cases, like spinach, where male and female flowers are produced on separate plants, the formation of male flowers is just as effective in causing senescence as is female flower production; even though the formation of male flowers requires much less in the way of nutrient consumption than the female flower, and fruit and seed development. In many trees, flowering occurs early in a growing season, sometimes even before leaf buds burst; not just before leaf fall, in other words.

The linkage between reproductive organs and senescence in monocarpic plants is very variable, anyway. In some species (pulses such as peas, soybeans, and beans are good examples) only the appearance of the seeds, not when flowers are formed, triggers senescence. In others, senescence is triggered by the flowers themselves or, sometimes, even by just the right conditions for flower production, without actual flower formation itself. This is the case in plants like spinach and hemp in which senescence occurs following flowering of the separate male and female plants. In a few other cases, removal of flowers will not delay but, rather, stimulates plant senescence, this being the case with maize and pepper.

All of these examples indicate that the signals, and maybe even the reasons, for senescence vary widely, at least among the many kinds of plants that flower once and then die.

While the redirection of nutrients seems an inadequate explanation of why plants enter senescence, that growth substances are involved seems, on the other hand, quite likely. Decrease in the supply of growth substances, specifically cytokinins (see chapter 7) from the roots, is partially responsible for the start of senescence, in leaves at least. Conversely, artificially maintaining cytokinin supply to leaves can delay leaf fall.

Unlike the cytokinins, other plant growth substances, especially ethylene but also sometimes ABA, accelerate senescence and death. In some fruits, ethylene production speeds up ripening and eventual fruit-drop enormously (see chapter 7); in flowers, the commonest effect of ethylene is the onset of wilting, followed by the fading of color, the export of nutrients, withering, and finally flower fall. But whether there is such a thing as a designated, as yet undiscovered, growth substance in plants which specifically triggers senescence remains an unanswered question.

Whatever is responsible for triggering senescence, there is an advantage to having regulated programs for the death and shedding of senescent leaves, flowers, and fruits. In the case of fruits, shedding is obviously important; after all, fruits contain the seeds which need to be dispersed. In the case of old flowers, the probability is that shedding involves removal of a useless organ that might otherwise act as a potential entry point for infection. Regular removal of old leaves as well allows renewal of the plant by new leaves that are not then hindered in their deployment of nutrients by the previous, diseased, half-eaten set.

Senescence could also be seen as an aid to survival rather than a prelude to death in some instances. When senescence occurs in any tissue of the plant, the breakdown and redistribution of carbohydrates, lipids, proteins, and other such large molecules mobilizes valuable nutrients which can be used over again. This nutrient economy helps certain plants, like forest trees for example, to survive in soils that are not very fertile. In this way, valuable nutrients can be recycled within the plant. In the case of mineral nutrients, this means that new supplies do not continually have to be scavenged from the surrounding, meager soil. Leaves that are shed in the fall from deciduous trees in temperate climates probably would have little chance of withstanding cold winters, so their loss, preceded by nutrient salvage, probably increases the chances of survival and productivity of individual perennial plants.

The kind of life strategy adopted by any living thing depends very much on its environment. Whether it is better to breed once or many times depends on the chance of survival of the offspring. In the environments in which adult mortality rates are high, for example, those in which the climate is severe, or in which predators are numerous, breeding just once may be best; less severe or hostile environments may favor a multiple breeding strategy.

Thus, many factors combine to determine the life history strategy each species of plant adopts and, therefore, how individuals of a species, finally, get dead, in whole or in part.

Epilogue

Green plants dominate our planet yet are often taken for granted. For many people, they are merely the passive aspect of a beautiful landscape, the 'backdrop' against which animals exist. For others, they are essential to their lives but are there simply to be exploited for food, fodder, fuel, furniture, clothing, transport, recreation, health purposes, and protection without thought being given to their unique qualities as living things in their own right.

In the preceding 17 chapters, I have attempted to provide some insights into the very different world of green plants. Their lives are lived at a different pace from ours, which may be one reason why we so often forget they are living organisms capable of doing so many of the things we also do. Like us, but in their own way, they can see, they can count, they can communicate with one another, they can be sensitive to the slightest touch, and they can tell the time with considerable precision. But they accomplish all of these things on a different time-scale from most animals. Their very slowness deceives us into believing that they do not do much at all.

We should not forget, however, that green plants along with some protists (see chapter 2) are unique among all organisms on earth in that they alone have the means to use light as a source of energy. The very substance which renders them green, the pigment chlorophyll, puts them in the position of being the very foundation of our biosphere. They, alone, sustain all of the remainder of life on this planet with the exception of a small number of curious, fascinating organisms which derive the energy they need from volcanic vents, rather than from the sun. Without green plants, the earth would be a stark, largely lifeless place.

Inwardly, green plants are highly active, wonderfully complex chemical factories. The sugars they form with the aid of light energy in photosynthesis are just the first few of many thousands of chemicals plants manufacture for their own use. After all, plants must grapple with many of the same problems as animals. First and foremost, they must feed and water themselves in order to sustain their growth. They must reproduce to ensure their survival into the next generation. In order to survive at all, they must fight their enemies which are numerous given that they are the primary source of food upon which all other organisms depend. They have to struggle with their neighbors for space in which to live and to gather nutrients for themselves. They must combat the elements to which they are exposed at all times given their relative immobility. The myriad of chemicals they produce serve as essential agents in one or more of these functions.

Plants are, in many ways, much more successful organisms than animals. They preceded animals onto the land, they thrive in places where animals find it difficult or impossible to survive, and they can grow bigger and live longer than animals. And, in the final analysis, as has been said already, animals are totally dependent on them.

All of these features of green plants are reasons enough to spend time learning something of what makes them unique, why they are so successful, where they live and . . . how they work!

Bibliography

Barz, W., Bless, W., Borger-Papendorf, G., Gunia, W., Mackenbrock, U., Meier, D., Otto, Ch. and Super, E. Phytoalexins as part of induced defence reactions in plants: their elicitation, function and metabolism. In *Bioactive compounds from plants,* Ciba Foundation Symposium 154, Wiley and Sons, 1990, pp. 140–156.

Bazzaz, F.A. and Fajer, E.D. Plant life in a CO_2-rich world, *Scientific American,* January, 1992, pp. 68–74.

Bender, D.A. *Amino acid metabolism,* Wiley and Sons, 1975.

Björkman, O. and Berry, J. High-efficiency photosynthesis, *Scientific American,* October, 1973, pp. 80–93.

Bonner, J. Chemical sociology among the plants. In *Plant life,* Scientific American Publication, 1957, pp. 156–162.

Brill, W.J. Biological nitrogen fixation, *Scientific American,* March, 1977, pp. 68–81.

Canny, M.J. *Phloem translocation,* Cambridge University Press, 1973.

Cox, P.A. Ethnopharmacology and the search for new drugs. In *Bioactive compounds from plants,* Ciba Foundation Symposium 154, Wiley and Sons, 1990, pp. 40–55.

Crafts, A.S. *Translocation in plants,* Holt, Rinehart and Winston, 1961.

Crawley, M.J. (ed.) *Plant Ecology,* Blackwell, 1986.

Darwin, C. *The origin of species by means of natural selection,* Murray, 1860.

Delph, L.F. and Lively, C.M. Pollinator visits to floral colour phases of *Fuchsia excorticata*, *New Zealand Journal of Zoology,* **12**: 599–603, 1985.

Dostrovsky, I. Chemical fuels from the sun, *Scientific American,* December, 1991, pp. 102–107.

Epstein, E. *Mineral nutrition of plants: principles and perspectives,* Wiley and Sons, 1972.

Flowers, T.J. and Yeo, A.R. *Solute transport in plants,* Blackie Academic and Professional, 1992.

Gabriel, M.L. and Fogel, S. (eds.) *Great experiments in biology,* Prentice-Hall, 1955.

Galston, A.W. and Davies P.J. *Control mechanisms in plant development,* Prentice-Hall, 1970.

Gilbert, F.A. *Mineral nutrition and the balance of life,* University of Oklahoma Press, 1957.

Gilbert, L.E. and Raven, P.H. (eds.) *Coevolution of animals and plants,* University of Texas Press, 1975.

Goodwin, T.W. and Mercer, E.I. *Introduction to plant biochemistry,* 2nd edition, Pergamon Press, 1983.

Greulach, V.A. The rise of water in plants. In *Plant life,* Scientific American Publication, Simon and Schuster, 1957, pp. 119–125.

Hales, S. The movement of liquids in plants studied by rigorous quantitative procedures. In *The origin and growth of biology,* A. Rook (ed.), Penguin, 1964, pp. 202–214.

Harbourne, J.B. *Introduction to ecological biochemistry,* 3rd edition, Academic Press, 1988.

Harbourne, J.B. Role of secondary metabolites in chemical defence mechanisms in plants. In *Bioactive compounds in plants,* Ciba Foundation Symposium 154, Wiley and Sons, 1990, pp. 126–139.

Harlan, J.R. The plants and animals that nourish man, *Scientific American,* September, 1976, pp. 89–97.

Hewitt, E.J. and Smith, T.A. *Plant mineral nutrition,* The English Universities Press Ltd, London, 1975.

Hopkins, W.G. *Introduction to plant physiology,* Wiley and Sons, 1995.

Huxley, A. *Plant and Planet,* Penguin, 1978.

Ingen-Housz, J. Experiments upon vegetables. In *Great experiments in biology,* M.L. Gabriel and S. Fogel (eds.), Prentice-Hall, 1955, pp. 158–161.

Janick, J., Noller, C.H. and Rhykerd, C.L. The cycles of plant and animal nutrition, *Scientific American,* September, 1976, pp. 75–86.

Jensen, W.A. *The plant cell,* Wadsworth, USA, 1964.

Jensen, W.A. and Salisbury, F.B. *Botany*, 2nd edition, Wadsworth, 1984.

Kendrick, R.E. and Frankland, B. *Phytochrome and plant growth,* Edward Arnold, 1976.

King, J. *The genetic basis of plant physiological processes,* Oxford University Press, 1991.

Klein, R.M. *The green world: an introduction to plants and people*, 2nd edition, Harper and Row, 1987.

Leopold, A.C. and Kriedemann, P.E. *Plant growth and development,* 2nd edition, McGraw-Hill, 1975.

Lesham, Y.Y., Halevy, A.H. and Frenkel, C. *Processes and control of plant senescence,* Elsevier, 1986.

Lewington, A. *Plants for people,* Oxford University Press, 1990.

Lewis, O.A.M. *Plants and nitrogen,* Edward Arnold, 1986.

Medina, J. *The clock of ages,* Cambridge University Press, 1996.

Naylor, A.W. The control of flowering. In *Plant life,* Scientific American Publication, Simon and Schuster, 1957, pp. 14–26.

Newman, E.I. Allelopathy: adaptation or accident? In *Biochemical aspects of plant and animal co-evolution,* J.B. Harborne (ed.), Academic Press, 1978.

Postgate, J.R. *The fundamentals of nitrogen fixation,* Cambridge University Press, 1982.

Priestley, J. Observations on different kinds of air. In *Great experiments in biology,* M.L. Gabriel and S. Fogel (eds.), Prentice-Hall, 1955, pp. 155–157.

Richardson, M. *Translocation in plants,* Edward Arnold, 1968.

Rosenthal, G.A. The chemical defenses of higher plants, *Scientific American,* January, 1986, pp. 94–99.

Sage, L.C. *Pigment of the imagination. A history of phytochrome research,* Academic Press, 1992.

Salisbury, F.B. and Ross, C.W. *Plant Physiology,* 4th edition, Wadsworth, 1992.

Saussure, N.Th. de On the influence of carbonic acid gas on mature plants. In *Great experiments in biology,* M.L. Gabriel and S. Fogel (eds.), Prentice-Hall, 1955, 161–165.

Schery, R.W. *Plants for man*, 2nd edition, Prentice-Hall, 1972.

Schocken, V. The auxins. In *Plant life,* Scientific American Publication, Simon and Schuster, 1957, pp. 3–13.

Skelton, P. (ed.) *Evolution: a biological and palaeontological approach,* Addison-Wesley Publishing Co., 1993.

Stiles, W. and Leach, W. *Respiration in plants,* 3rd edition, Methuen, 1952.

Sutcliffe, J. Plants and water. *Studies in Biology No. 14*, Edward Arnold, 1968.

Sutcliffe, J.F. and Baker, D.A. *Plants and mineral salts,* Edward Arnold, 1981.

Taiz, L. and Zeiger, E. *Plant Physiology,* Benjamin/Cummings, 1991.

Thimann, K.V. Autumn colors. In *Plant life,* Scientific American Publication, Simon and Schuster, 1957, pp. 100–109.

Thimann, K.V. *Senescence in plants,* C.R.C. Press, 1980.

Thomas, M., Ranson, S.L. and Richardson, J.A. *Plant Physiology,* 5th edition, Longman, 1973.

Tribe, M. and Whittaker, P. Chloroplasts and mitochondria. *Studies in Biology No. 31,* Edward Arnold, 1972.

Tudge, C. *The environment of life,* Oxford University Press, 1988.

Van Helmont, J.B. By experiment, that all vegetable matter is totally and materially of water alone. In *Great experiments in biology,* M.L. Gabriel and S. Fogel (eds.), Prentice-Hall, 1955.

Villiers, T.A. *Dormancy and the survival of plants,* Edward Arnold, 1975.

Weiss, M.R. Floral colour changes as cues for pollinators, *Nature* **354**: 227–229, 1991.

Wheelwright, N.T. and Janson, C.H. Colors of fruit displays of bird-dispersed plants in two tropical forests, *American Naturalist* **126**: 777–799, 1985.

Wildon, D.C., Thain, J.F., Minchin, P.E.H., Gubb, I.R., Reilly, A.J., Skipper, Y.D., Doherty, H.M., O'Donnell, P.J. and Bowles, D.J. Electrical signalling and systemic proteinase inhibitor induction in the wounded plant, *Nature*, **360**: 62–65, 1992.

Wilkins, M.B. *The physiology of plant growth and development,* McGraw-Hill, 1969.

Woolhouse, H.W. Ageing processes in higher plants, *Oxford Biology Reader No. 30*, J.J. Head and O.E. Lowenstein (eds.), Oxford University Press, 1972.

Index